BEAM STABILITY AND NONLINEAR DYNAMICS

The Santa Barbara Symposium on *Beam Stability and Nonlinear Dynamics*, brought together leaders from accelerator physics communities, mathematics, and other fields of physics to deal with some of the fundamental theoretical problems of accelerator physics. The group picture shows some of the participants. (See additional group pictures in the back).

BEAM STABILITY AND NONLINEAR DYNAMICS

Santa Barbara, California   December 1996

EDITOR
Zohreh Parsa
Brookhaven National Laboratory

AIP CONFERENCE
PROCEEDINGS 405

American Institute of Physics   Woodbury, New York

Authorization to photocopy items for internal or personal use, beyond the free copying permitted under the 1978 U.S. Copyright Law (see statement below), is granted by the American Institute of Physics for users registered with the Copyright Clearance Center (CCC) Transactional Reporting Service, provided that the base fee of $10.00 per copy is paid directly to CCC, 222 Rosewood Drive, Danvers, MA 01923. For those organizations that have been granted a photocopy license by CCC, a separate system of payment has been arranged. The fee code for users of the Transactional Reporting Service is: 1-56396-731-6/ 97 /$10.00.

© 1997 American Institute of Physics

Individual readers of this volume and nonprofit libraries, acting for them, are permitted to make fair use of the material in it, such as copying an article for use in teaching or research. Permission is granted to quote from this volume in scientific work with the customary acknowledgment of the source. To reprint a figure, table, or other excerpt requires the consent of one of the original authors and notification to AIP. Republication or systematic or multiple reproduction of any material in this volume is permitted only under license from AIP. Address inquiries to Office of Rights and Permissions, 500 Sunnyside Boulevard, Woodbury, NY 11797-2999; phone: 516-576-2268; fax: 516-576-2499; e-mail: rights@aip.org.

L.C. Catalog Card No. 97-74676
ISBN 1-56396-731-6
ISSN 0094-243X
DOE CONF- 961208

Printed in the United States of America

Contents

Preface vii

From Taylor Series to Taylor Models 1
    M. Berz

Numerical Evaluation of Long-Term Stability 25
    M. Giovannozzi, W. Scandale, and E. Todesco

Convergence of a Fourier-Spline Representation for the Full-Term Map Generator 41
    R. L. Warnock and J. A. Ellison

Algorithms for the Treatment and Analysis of Spin Dynamics in SU(2) and SO(3) 55
    V. Balandin, M. Berz, and N. Golubeva

Normal Form Analysis of the LHC Dynamic Aperture 63
    F. Schmidt and E. Todesco

Neutralised Beams: Landau Damping in System with Strong Nonlinearity 75
    P. Zenkevich and E. Mustafin

Nonlinear Dynamics of Accelerator via Wavelet Approach 87
    A. N. Fedorova and M. G. Zeitlin

Some Problems of Nonlinear Aberration Correction 103
    S. N. Adrianov

On the Mechanism of the Saw-Tooth Instability 117
    S. Heifets

Integrable Cases in Nonlinear Betatron Motion 133
    J. Hagel

Nonlinear Structures Near the Boundary of Marginal Stability 149
    T. A. Davydova and A. Yu. Pankin

Kinetics of Muon Longitudinal Cooling 165
    Z. Parsa and P. Zenkevich

Decoherence and Wakefield Effects in One-Turn Map Measurement 173
    C.-xi Wang and J. Irwin

Application of Moments Method to Dynamics of Muon Cooling System 183
    Z. Parsa and P. Zenkevich

On 2-D Electron Cloud Dynamics in High-Current Plasma Lens for Ion Beam Focusing 189
    A. A. Goncharov, I. V. Litovko, N. Onishchenko, and V. F. Zadorozhny

Estimates for Long-Term Stability for the LHC 201
    M. Böge and F. Schmidt

Halo Control, Beam Matching, and New Dynamical Variables for Beam Distributions 211
    W. Lysenko and Z. Parsa

Schedule 223
List of Participants 227
Author Index 237

PREFACE

This Preface includes an introduction and summary of a Symposium on "Beam Stability and Nonlinear Dynamics" which was held December 3–5, 1996 at the Institute for Theoretical Physics (ITP) in Santa Barbara. This was the third symposium[1] hosted by the ITP and supported by its sponsor the National Science Foundation, as part of the 1st US long term accelerator research program, on "New Ideas for Particle Accelerators." The long term program and symposia were organized and coordinated by Dr. Zohreh Parsa of Brookhaven National Laboratory/ITP.

The purpose of this symposium was to deal with some of the fundamental theoretical problems of accelerator physics by bringing together leaders from accelerator physics communities, mathematics, and other fields of physics. The focus was on nonlinear dynamics and beam stability. The symposium began with some defining talks on relevant mathematical topics such as single-particle Hamiltonian dynamics, chaos, and new ideas in symplectic integrators. The physics topics included single-particle and many-particle dynamics. These topics concern circular accelerators in which particles circulate for a very large number of turns as well as linear accelerators where space charge and wakefields induced in accelerating cavities play a strong role.

A major question is to determine the best model for numerical simulations in order to accurately reproduce behavior of beams in real accelerators and to predict long-term or long distance stability. Comparison with experiment is recognized as an important tool in improving models.

Straight-forward tracking using linear elements and thin-lens multipoles to preserve symplecticity is the basic tool for studying single-particle dynamics and stability in large circular accelerators such as the Large Hadron Collider (LHC), which was recently approved for construction at CERN. Ideas have been aimed at improving the computation time and/or in improving analysis of the results. Symplectification of Taylor maps are used since truncation of expansion maps leads to maps that are not symplectic. The concept of jolt factorization makes it possible to obtain a symplectic truncated expansion. But if the nonlinearity is too large, as is usually the case near the onset of unstable motion, map predictions fail. This raises the difficult question of the applicability of a complete-turn map to a large accelerator. The expansion of such maps is laborious for phase-space dimensions larger than four. Another related development is the use of Taylor's models with additional functions which bound the initial function from above and below. Application of this concept to maps led to the development of an arithmetic, which applies to both the polynomial and the remainders, termed Remainder Differential Algebra. This should provide information on the accuracy of the map description. Symplecticity is ensured if we used the Hamiltonian formalism and action-angle variables. In this approach, the map over one turn or a fraction of turn can be computed by solving algebraic equations related to canonical transformations which are in implicit form. This is done for the non-periodic solutions of the generating function equation by using Newton iterations and approximation in Fourier series and B-spline functions.

Interesting results of numerous trackings and analyses (including those developed during our ITP workshop) were presented for the LHC. Different methods for estimating the dynamic aperture were tested, first using the Henon map. Early indicators such as the Lyapunov criterion, frequency map analysis, and variation of tunes have been used with tracking over an increasing number of turns. A new conjecture combining the result of the KAM theorem with the Nkhoroshev estimate

---

[1] In addition to this conference, a week long symposium was held on "New Modes of Particle Acceleration—Techniques and Sources" August 19–23, 1996. Some of the highlights of that meeting included Novel Modes of laser, plasma, wakefield accelerations, techniques and power sources. A second symposium a week long was held on "Future High Energy Colliders" October 21–25, 1996. Some of the highlights of that meeting included discussions on the future direction of high energy physics by bringing together leaders from the theoretical, experimental and accelerator physics communities. The perspectives that were presented at the symposium on the state and future of high energy physics are vital ingredients in the continuing discussion of how the U.S. High Energy Physics community should best marshal its national scientific resources while continuing a high level of international collaboration.

predicts that the dynamic aperture depends on the inverse logarithm of the number of turns. There seemed to be a good agreement between the predictions of the early indicators and the result of the conjecture extrapolated to a very large number of revolutions. This gives an increased confidence in the numerical predictions, to within 10 or 20% of the actual value as supported by measurements on existing accelerators.

Particular examples of stability analysis were presented, such as a Hamiltonian system with a quartic potential and the three-body problem in celestial mechanics. Linearization around a periodic solution in the first case and around a Lagrangian fixed point in the second, provides a monodromic matrix which give information on the stability. For the three-body problem, developments to second order allows us to solve the equation of motion near resonance. Also presented was the idea to apply to accelerator dynamics the wavelet analysis of Hamiltonian systems. New Applications of moment method to study kinetics and dynamics of muon cooling systems and the use of moments and new variables for Beam matching and Halo control were also very interesting.

Among the other subjects treated were spin dynamics, nonlinear aberration correction including space charge aberrations, collective effects in the LHC, sawtooth instability, and Landau damping in the presence of strong nonlinearity. There were other presentations concerning plasma physics effects relevant to accelerators, and the peculiar effect of beam echos that has recently been observed for the first time in an existing accelerator with echo times as long as one to two minutes. Numerical tools for studying multibunch instability in linear accelerators with strong wakefields were presented, together with a statistical method of analysis of wakefield effects on emittance growth, based on beamline response coefficients.

The conference ended with a unique discussion session in which participants presented and clarified their views on outstanding problems and topics presented at the symposium. This international forum has provided new and valuable input for future developments in this field.

At the request of participants, an additional post-conference session was held on December 6, 1996. The write-ups of some of those presentations and topics are also included in these proceedings. Our special thanks to Mrs. Storm and her staff for making the arrangements for the sessions.

I would like to thank all the authors for providing the write-ups of their talks.

In addition, for one reason or another, several speakers were not able to provide the write-ups of their talks. Nevertheless, I thank them for their stimulating talks and participation in the symposium. In most cases, copies of the transparencies from their talks can be found in the report BNL-52525 and NSF-ITP-96-155i. These include presentations by Drs. J. Meiss "Single-Particle Hamiltonian Dynamics," J. Marsden on "Symplectic Geometry, Maps, Integrators," J. Laskar on "Frequency Map Analysis—Theory and Experiment," F. Ruggiero on "Longitudinal Beam Echoes," R. Siemann on "Sawtooth Instability in the SLC Damping Rings," H. Yoshida on "Instability of Periodic Orbit and Non-integrability of Hamiltonian System," G. Guignard on "Stability of Beams in a High Frequency Linac with Strong Wakefields," etc.

For a complete list of the symposium presentations see the program schedule given in the Appendix.

I would like to thank the advisory committee, all speakers, conveners and participants for making the symposium a unique and stimulating experience. I also thank W. Lysenko, R. Warnock, P. Zenkevich and other participants for providing the photos, and special thanks to the ITP Director, Manager, and staff for providing a beautiful setting and making sure the meeting ran smoothly.

Zohreh Parsa
Chairperson, Symposium
Institute for Theoretical Physics,
UCSB, Santa Barbara, California
Brookhaven National Laboratory, Upton, New York

From Taylor Series to Taylor Models

Martin Berz

Department of Physics and Astronomy and
National Superconducting Cyclotron Laboratory
Michigan State University
East Lansing, MI 48824

Abstract

An overview of the background of Taylor series methods and the utilization of the differential algebraic structure is given, and various associated techniques are reviewed. The conventional Taylor methods are extended to allow for a rigorous treatment of bounds for the remainder of the expansion in a similarly universal way. Utilizing differential algebraic and functional analytic arguments on the set of Taylor models, arbitrary order integrators with rigorous remainder treatment are developed. The integrators can meet pre-specified accuracy requirements in a mathematically strict way, and are a stepping stone towards fully rigorous estimates of stability of repetitive systems.

INTRODUCTION

The year 1996 marks the tenth anniversary[1] of the introduction of the differential algebraic approach[2,3] into the study of beam dynamics. It took the computation of Taylor maps

$$\vec{z}_f = \mathcal{M}(\vec{z}_i) \qquad (1)$$

of dynamical systems from the then customary third[4,5,6,7] or fifth order[8] all the way to arbitrary order in a unified and straightforward way. The Taylor maps have many applications, as many of the **physical quantities** that are encountered in practice are more or less **directly connected to Taylor coefficients**. Since its introduction, the method has been widely utilized in a large number of new map codes.[9,10,11,12,13,14,15,16,17,18,19]

The basic idea behind the method is to bring the treatment of **functions** to the computer in a similar way as the treatment of **numbers**. In a strict sense, neither functions (for example, C^∞) nor numbers (for example, the reals R) can be treated on a computer, since neither of them can be represented with the finite amount of information that can be stored on computers (after all, a real number is an equivalence class of bounded Cauchy sequences of rational numbers).

However, from the early days of computers we are used to dealing with numbers by **extracting information deemed relevant**, which in practice usually means the approximation by **floating point numbers** with finitely many digits. In a formal sense this is possible since for every one of the operations on real numbers, like addition and multiplication, we can craft an **adjoint** operation on the floating point numbers such that the following diagram commutes:

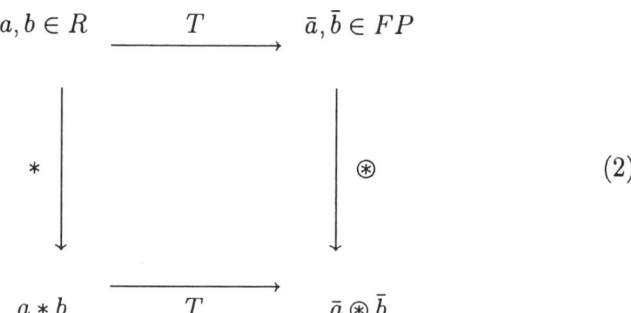

$$
\begin{array}{ccc}
a, b \in R & \xrightarrow{T} & \bar{a}, \bar{b} \in FP \\
* \downarrow & & \downarrow \circledast \\
a * b & \xrightarrow{T} & \bar{a} \circledast \bar{b}
\end{array}
\qquad (2)
$$

Of course, much to the chagrin of those doing numerics, in reality the diagrams commute only "approximately", which typically makes the errors grow over time.

The approximate character of these arguments can be removed by representing a real not by one floating point number, but rather by an **interval** of floating point numbers providing a rigorous upper and lower bound. By rounding operations down for lower bounds and up for upper bounds, rigorous bounds can be found for sums and products, and adjoint operations can be made such that the above diagram commutes exactly. In practice, while always maintaining rigor, the method sometimes becomes rather pessimistic, as over time the intervals often have a tendency to grow.

Historically, the treatment of **functions** in numerics has been done based on the treatment of **numbers**; and as a result, virtually all classical numerical algorithms are based on the mere evaluation of functions at specific points. As a consequence, numerical methods for differentiation, which are so relevant for the computation of Taylor representations of the map (1), are very cumbersome and prone to inaccuracies because of cancellation of digits, and not useful in practice for our purposes.

The success of the new methods is based on the observation that it is possible to extract more information about a function than its mere values. Indeed, considering the commuting diagram in eq. (2), one can demand the operation T to be the extraction of the Taylor coefficients of a pre-specified order n of the function. In mathematical terms, T is an equivalence relation, and the application of T corresponds to the transition from the function to the **equivalence class** comprising all those functions with identical Taylor expansion to order n.

Since Taylor coefficients of order n for sums and products of functions as well as scalar products with reals can be computed from those of the summands and factors, it is clear that the diagram can be made to commute; indeed, except for the underlying inaccuracy of the floating point arithmetic, it will even commute exactly. In mathematical terms, this means that the set of equivalence classes of functions can be endowed with well-defined operations, leading to the so-called **Truncated Power Series Algebra**[1].[2]

This fact was realized in the first paper on the subject,[2] which led to a method to extract maps to any desired order from a computer algorithm that integrates orbits numerically. Similar to the need for algorithms within floating point arithmetic, the development of **algorithms for functions** followed, including methods to perform composition of functions, to invert them, to solve nonlinear systems explicitly, and to introduce the treatment of common elementary functions.[20,21]

However, very soon afterwards it became apparent[22,3] that this only represents a half-way point, and one should **proceed beyond mere arithmetic operations** on function spaces of addition and multiplication and consider their **analytic operations of differentiation and integration**. This resulted in the recognition of the underlying **differential algebraic structure** and its practical exploitation, based on the commuting diagrams for addition, multiplication, and differentiation and their inverses:

$$
\begin{array}{ccccccc}
f,g & \xrightarrow{T} & F,G & f,g & \xrightarrow{T} & F,G \\
{\scriptstyle +,-}\downarrow & & \downarrow{\scriptstyle \oplus,\ominus} & {\scriptstyle \cdot,/}\downarrow & & \downarrow{\scriptstyle \odot,\oslash} \\
f\pm g & \xrightarrow{T} & F{\oplus\atop\ominus}G & f\,\dot{}\,/g & \xrightarrow{T} & F{\odot\atop\oslash}G
\end{array}
$$

$$
\begin{array}{ccc}
f & \xrightarrow{T} & F \\
{\scriptstyle \partial,\partial^{-1}}\downarrow & & \downarrow{\scriptstyle \partial_\bigcirc,\partial_\bigcirc^{-1}} \\
\partial f,\partial^{-1}f & \xrightarrow{T} & \partial_\bigcirc F,\partial_\bigcirc^{-1}F
\end{array}
\qquad (3)
$$

In passing we note that in order to avoid loss of order, in practice the derivations have the form $\partial = h \cdot d/dx_i$, where h is a function with $h(0) = 0$. As a first consequence, it allowed to construct integration techniques to any order that for a given accuracy demand are substantially faster than conventional methods.[21] Subsequently, it was realized that the differential algebraic operations are useful for a whole variety of other questions connected to the analytic properties of the transfer map.[20] It was possible to determine arbitrary order **generating function** representations of maps[23,21]; factorizations into **Lie operators**[24] could be carried out for the first time to arbitrary order[21]; **normal form methods**[25,26] could be performed to arbitrary order.[27,21] And last but not least, the complicated **PDEs** for the fields and potentials stemming from the representation of Maxwell's equations in particle optical coordinates could be solved to any order in finitely many steps.

Of course the question of what constitutes "information deemed relevant" for functions does not necessarily have a unique answer. Formula manipulators, for example, attack the problem from a different perspective by attempting to algebraically express functions in terms of certain elementary functions linked by algebraic operations and composition. In practice the

Achilles heel of this approach is the complexity that such representations can take after only a few operations. But compared to the mere Taylor expansion, they have the advantage of rigorously representing the function under consideration. Below we will show how such rigor can be maintained without the computational expense of formula manipulation by a suitable augmentation of the Taylor approach.

TRUNCATED POWER SERIES, DIFFERENTIALS, DIFFERENTIAL ALGEBRAS, AND AUTOMATIC DIFFERENTIATION

Before proceeding further, it seems to be worthwhile to put into perspective a variety of different concepts that were introduced to the field in connection with the above developments. We do this for a dual purpose: on the one hand we hope to alleviate some of the confusion in the field resulting from an overly casual and often improper use of terminology; and on the other hand, we want to try to provide a summary of various useful techniques outside the field. Furthermore we lay the groundwork for the further development in the next sections, in which differential algebraic techniques will be applied to a new set of objects.

The first and simplest structure that was introduced[12] is **TPSA**, the truncated power series algebra. This is the structure that results when the equivalence classes of functions are endowed with arithmetic such that the diagrams in eq. (2) commute for the basic operations of addition, multiplication, and scalar multiplication. Addition and scalar multiplication lead to a **vector space**, and the multiplication operation turns it into a commutative **algebra**. In many respects, together with the polynomial algebras, this structure is an archetypal non-trivial algebra, and in fact it can be embedded into many larger and more interesting algebras.

It is easy to see that the **TPSA** can be equipped with an order, and then contains **differentials**, i.e. infinitely small numbers. This fact triggered the study of such nonarchimedean structures in more detail, and led to the introduction of a foundation of analysis[28,29,30] on a larger and for such purposes much more useful structure, the Levi-Civita field. It turned out that the Levi-Civita field is the smallest nonarchimedean extension of the real numbers that is algebraically and Cauchy complete, and many of the basic theorems of calculus can be proved in a similar way as in R. Furthermore, concepts like Delta functions and the idea of derivatives as differential quotients can be formulated rigorously and integrated seamlessly into the theory. On the practical end, based on the latter concept, there are also several improvements regarding methods of computational differentiation.[31,32]

As alluded to in the last section, the power of TPSA can be enhanced by the introduction of derivations ∂ and their inverses, corresponding to the differentiation and integration on the space of functions. It was mentioned that the resulting structure, a **Differential Algebra**, allowed the direct treatment of many questions connected with differentiation and integration of functions, including the solution of the ODEs describing the motion and PDEs describing the fields, as well as the determination of generating functions and Lie factorizations to arbitrary order.[21]

These applications follow in the vein of other applications of differential algebras, the study of which became important connected to the question of **solving analytic problems with algebraic means.** Among others, this work was initiated in a serious fashion by Liouville[33] connected to the problem of integration of functions and differential equations in finite terms. It was then significantly enhanced by Ritt,[34] who provided a rather complete algebraic theory of the solution of differential equations that are polynomials of the functions and their derivatives and that have meromorphic coefficients. Further development in the field is due to Kolchin[35] and, already with an eye on the algorithmic aspect, to Risch.[36,37,38]

Nowadays the methods form the basis of many algorithms in modern formula manipulators, where the treatment of differential equations and quadrature problems calls for the solution of analytic problems with algebraic means. Other important current work relying on differential algebraic methods is the practical study of differential equations under algebraic constraints, so-called differential algebraic equations.[39] Many of the recent developments will be covered in a forthcoming special issue on Differential Equations and Differential Algebra of the Journal of Symbolic Computation.

The final concept that is somewhat connected to our methods and worth to be studied is the technique of **automatic differentiation.**[40,41,42] The purpose of this discipline is the automated transformation of existing code in such a way that derivatives of functional relationships between variables are calculated along with the original code. Besides the significantly increased computational accuracy compared to numerical differentiation, a striking advantage of this approach is the fact that in the so-called reverse mode it is actually possible in principle to calculate gradients in v variables in a **fixed amount of effort**; independent of v, in the optimal case the entire gradient can be obtained with a cost equalling only about five times the cost of the evaluation of the original functions, in stark contrast to numerical differentiation requiring $(v+1)$ times the original cost.

In practice, automatic differentiation is almost exclusively **first order**, and as such is not directly useful for our purposes. One reason for this situation is connected to the fact that conventional numerical algorithms avoid higher derivatives as much as possible because of the well-known difficulties when trying to obtain them via numerical differentiation, which for a long time represented the only available approach. On the other hand, the above mentioned savings that are possible for linear derivatives are much harder to obtain in the same way for higher orders.

In passing it may be worthwhile to note that contrary to what may be expected at first sight, the automatic differentiation community is not quite readily embracing the computational simplifications of modern object oriented techniques. Aside from the fact that the problem usually involves the need of making adjustments to existing code and the fact that the reverse approach requires code re-structuring and not just operator overloading, it has often proven difficult to obtain competitive computational performance.

Altogether, the challenge in automatic differentiation is more **reminiscent of sparse matrix techniques** for management and manipulation of Jacobians than of a power series technique. It is perhaps also worth mentioning that because of the need for code re-structuring in order to obtain performance, there is a certain reluctance in the community towards the use of the word "automatic". Mostly in order to avoid the impression of making false promises, the technique recently likes to refer to itself as **computational differentiation**.

Only very recently are other groups in computational differentiation picking up at least on second order,[43] but so far the only software for derivatives beyond order two listed in the automatic differentiation tool compendium[44] is in fact the package DAFOR[45,46,47] consisting of the FORTRAN precompiler DAPRE and the arbitrary order DA package that is also used as the power series engine in the code COSY INFINITY.

It is the author's hope that researcher in our field will in the future more seriously follow some of these leads into neighboring disciplines, and that he would more frequently meet some of his colleagues at the many conferences of these fields. On the one hand, there are a variety of interesting techniques that may be borrowed; on the other hand, it is important to make the field of beam dynamics and its interesting problems more known in other communities.

THE TREATMENT OF REMAINDERS

Compared to techniques of formula manipulation and to other rigorous mathematical efforts on computers, the Taylor DA methods have the disadvantage that there is no way to make any statements about the remainder of Taylor's formula. It is our goal to extend the theory in such a way that it is possible to obtain rigorous bounds for the remainder terms. In this endeavour, we will have the demand to be fully mathematically rigorous in that no approximations are allowed. All this will be achieved by keeping the idea of providing commuting diagrams for elementary operations; however, the objects on which these operations are to be carried out are not mere truncated Taylor series any more, but rather new objects called **Taylor models.**

Furthermore, in order to keep the mathematical rigor for the solution of the differential equations defining the maps of the systems, we will derive a new method to perform integration. As in many other automated approaches for integration of functions and differential equations on computers, we will **utilize differential algebraic techniques** for this purpose. While in the conventional computation of Taylor maps, in principle also conventional integrators can be used (although the ones that come for free in the differential algebraic approach are usually superior in speed and accuracy), this is not the case here, and one is more or less forced to develop new techniques.

Our method will rely on an inclusion of the remainder term of a Taylor expansion in an interval. However, to quell misunderstandings from the beginning, it is important to note that our approach is **not equivalent to interval methods** that have been applied extensively for many types of verified calculations. The careful reader will realize that our method provides remainder bounds with an accuracy that does not scale merely linear with the domain interval, but rather as a high power of the domain interval; this feature is essential if high accuracy is required over an extended range of arguments, as is the case with the transfer map. Furthermore, it alleviates the so-called dependency problem, which among other things entails that extended conventional interval computations sometimes have a danger to "blow up" and yield rather pessimistic and sometimes even useless bounds.

COMPUTATION OF REMAINDER BOUNDS FOR FUNCTIONAL DEPENDENCIES

We begin our study of the rigorous computational treatment of the remainder with the definition of a **Taylor Model**. Let f be $C^{(n+1)}$ on $D_f \subset R^v$, and $\vec{B} = [a_1, b_1] \times ... \times [a_v, b_v] \subset D_f$ an interval box containing the point $\vec{x}_0$. Let T be the Taylor polynomial of f around the point $\vec{x}_0$. We call the interval I an nth order **Remainder Bound** of f on $\vec{B}$ if

$$f(\vec{x}) - T(\vec{x}) \in I \text{ for all } \vec{x} \in \vec{B}.$$

In this case, we call the pair (T, I) an nth order **Taylor Model** of f. It is clear that a given function f can have many different Taylor models, as with (T, I), also $(T, \bar{I})$ with $\bar{I} \supset I$ is a Taylor model. Furthermore, we see that low-order polynomials have trivial remainder bounds; since every polynomial of order not exceeding n agrees with its nth order Taylor polynomial, the interval $[0, 0]$ is a remainder bound.

For practical purposes, it is important that if the original interval box $\vec{B}$ decreases in size, then according to the various formulas of the Taylor remainder,[48] the remainder bounds can decrease in size with a power of $n+1$ and hence will become small quickly. In particular, this entails that the knowledge of a good Taylor model of a function on an interval box $\vec{B}$ allows a rather accurate estimate of the range of the function.

Now we want to study to what extent it is possible to define arithmetic operations $\oplus$, $\odot$, and $\partial_\bigcirc$ on Taylor models. In this case, the operation "T" that turns a function into its Taylor polynomial has to be replaced by the inclusion operation $\dot{\subset}$. So we must craft new **adjoint operations on Taylor models** that make the diagram

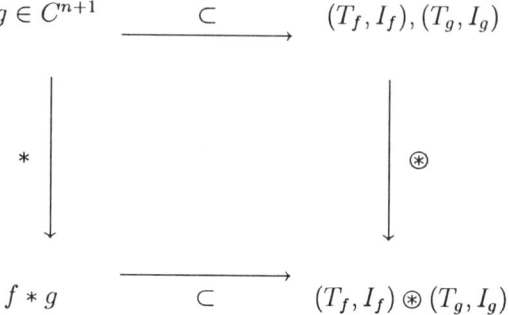

commute in a similar way as in the case of the Differential Algebra on Truncated Power Series in eq.(3).

Let (T_f, I_f) and (T_g, I_g) be nth order Taylor models of the functions f and g on the interval box $\vec{B}$. Clearly, the Taylor polynomial of $(f+g)$ is simply $T_f + T_g$; on the other hand, we know that on $\vec{B}$, $f(\vec{x}) \in T_f(\vec{x}) + I_f$ and $g(\vec{x}) \in T_g(\vec{x}) + I_g$. Then obviously,

$$(f+g)(\vec{x}) \in (T_f + T_g)(\vec{x}) + (I_f + I_g) \text{ for all } \vec{x} \in \vec{B},$$

and so $(T_f + T_g, I_f + I_g)$ is a Taylor model for $(f+g)$ on $\vec{B}$. And for practical purposes, it is also important to note that if I_f, I_g are "fine of order $\vec{B}^{n+1}$", i.e. their size scales with the size of $\vec{B}$ to the $(n+1)$st power, so is $I_{f+g} = I_f + I_g$. In the same way we see that $(T_f - T_g, I_f - I_g)$ is a Taylor model for $(f-g)$. So by simply defining

$$(T_f, I_f) \oplus (T_g, I_g) = (T_f + T_g, I_f + I_g),$$

we are able to close the commuting diagram for addition.

In order to study multiplication, let (T_f, I_f) and (T_g, I_g) be nth order Taylor models of the functions f and g on the interval box $\vec{B}$. As pointed out before, the Taylor polynomial $T_{f \cdot g}$ of $f \cdot g$ can then be obtained by multiplication of T_f and T_g and subtraction of the polynomial $\bar{T}_{f \cdot g}$ consisting of the terms whose order exceeds n. For any $\vec{x} \in \vec{B}$, there are values $e_f \in I_f$ and $e_g \in I_g$ such that $f(\vec{x}) = T_f(\vec{x}) + e_f$ and $g(\vec{x}) = T_g(\vec{x}) + e_g$. So we obtain

$$(f \cdot g)(\vec{x}) = (T_f(\vec{x}) + e_f) \cdot (T_g(\vec{x}) + e_g)$$
$$= T_f(\vec{x}) \cdot T_g(\vec{x}) + T_f(\vec{x}) \cdot e_g + T_g(\vec{x}) \cdot e_f + e_f \cdot e_g$$
$$= T_{f \cdot g}(\vec{x}) + \{\bar{T}_{f \cdot g}(\vec{x}) + T_f(\vec{x}) \cdot e_g + T_g(\vec{x}) \cdot e_f + e_f \cdot e_g\}.$$

The first term is the Taylor polynomial of $f \cdot g$. The term in curly brackets describes the behavior of the remainder; it is a polynomial in the $v + 2$ variables $(\vec{x}, e_f, e_g) \in \vec{B} \times I_f \times I_g$ and is denoted by $R(\vec{x}, e_f, e_g)$. So by bounding $R(\vec{x}, e_f, e_g)$[49,48] with an interval I_R, we are able close the diagram with the definition

$$(T_f, I_f) \odot (T_g, I_g) = (T_{f \cdot g}, I_R).$$

We note that the necessary computation of $T_{f \cdot g}$ from T_f and T_g is of course the standard multiplication within TPSA.

Besides providing the operations $\oplus$ and $\odot$ for Taylor models such that the diagrams in eq. (2) commute, there are a variety of other operations that have to be ported to the Taylor models, especially the intrinsic functions, the composition of functions, and several operations derived from these. For reasons of space, we have to restrict ourselves here to a referral to more detailed papers about the matter.[49,48]

Altogether, the operations $\oplus$ and $\odot$ enable us to determine mathematically rigorous bounds for the remainder of any function that can be represented on a computer, and is hence of great help for problems of optimization.[52] In itself, it also already useful for several problems in Beam Physics, in particular for the notoriously difficult bounding of approximate invariants of nonlinear motion.[50]

COMPUTATION OF REMAINDER BOUNDS FOR FLOWS OF DIFFERENTIAL EQUATIONS

Our goal is now to establish a Taylor model for the transfer map $\mathcal{M}(\vec{r}_0, t)$ in eq. (1), and thus in particular a rigorous bound for the remainder term of the flow of the differential equation describing the motion over a domain $(\vec{r}_{01}, \vec{r}_{02}) \times (t_0, t_2)$. As pointed out before, this need precludes us from the direct use of conventional numerical integrators, as they cannot provide rigorous bounds for the integration error but only approximate estimates. Rather, we have to start from scratch from the foundations of the theory of differential equations.

As a first step it is necessary to introduce the inverse derivation operation $\partial_\odot^{-1}$ on Taylor models. Given an n-th order Taylor model (P_n, I_n) of a function f, we can determine a Taylor model for the indefinite integral $\partial_i^{-1} f = \int f \, dx_i'$ with respect to variable i. The Taylor polynomial part is obviously just given by $\int P_{n-1}dx'_i$, and a remainder bound can be obtained as $(B(P_n - P_{n-1}) + I_n) \cdot B(x_i)$, where $B(x_i)$ is an interval bound for the variable x_i obtained from the range of definition of x_i, and $B(P_n - P_{n-1})$ is a bound for the part of P_n that is of exact order n. We thus define the operator $\partial^{-1}_{\bigcirc,i}$ on the space of Taylor models as

$$\partial^{-1}_{\bigcirc,i}(P_n, I_n) = \left(\int P_{n-1}dx'_i, \; (B(P_n - P_{n-1}) + I_n) \cdot B(x_i) \right). \quad (4)$$

The careful reader may perhaps wonder about the introduction of the operator $\partial_{\bigcirc,i}$; this is also possible, however at an additional effort, since from the knowledge of a remainder bound of a function, no conclusions can be drawn regarding a remainder bound for its derivative (for example, the function can oscillate very quickly inside even a narrow interval). With a further extension of the concept of Taylor models that also describes the asymptotic behavior of coefficients, this problem can be solved, but since it is not required for our purposes, we will not discuss the matter in detail here.

Schauder's Fixed Point Theorem

As is common for the application of functional analysis tools to the study of differential equations, we re-write the differential equation as an integral equation

$$\vec{r}(t) = \vec{r}_0 + \int_{t_0}^{t} \vec{F}(\vec{r}(t'), t') \, dt',$$

noting that the initial value problem has a (unique) solution if and only if the corresponding integral equation has a (unique) solution. Now we introduce the operator

$$A : \vec{C}^0[t_0, t_1] \to \vec{C}^0[t_0, t_1]$$

on the space of continuous functions from $[t_0, t_1]$ to R^n via

$$A\left(\vec{f}\right)(t) = \vec{r}_0 + \int_{t_0}^{t} \vec{F}(\vec{f}(t'), t') \, dt'; \quad (5)$$

so a general function $\vec{f}$ in $\vec{C}^0[t_0, t_1]$ is transformed into a new function in $\vec{C}^0[t_0, t_1]$ via the insertion into $\vec{F}$ and subsequent integration. Having introduced the operator A, the problem of finding a solution to the differential equation is reduced to a fixed-point problem

$$\vec{r} = A(\vec{r}).$$

It is common fare in the theory of differential equations to establish that Schauder's fixed point theorem asserts the existence of a solution of an ODE over the interval $[t_0, t_1]$ in case $\vec{F}$ is continous on $[t_0, t_1] \times R^n$ and bounded there. If $\vec{F}$ is even Lipschitz with respect to the first argument $\vec{f}$, then Banach's fixed point theorem even asserts a locally unique solution. However, in both cases the conventional results assert merely the existence of a solution and do not provide details about its range.

We will now apply Schauder's fixed point theorem[51] in a different way to rigorously obtain a Taylor Model for the flow.

Theorem (Schauder): *Let A be a continous operator on the Banach Space X. Let $M \subset X$ be compact and convex, and let $A(M) \subset M$. Then A has a fixed point in M, i.e. there is an $\vec{r} \in M$ such that $A(\vec{r}) = \vec{r}$.*

One should be reminded that the fixed point is not necessarily unique (for example, the identity map on M has every element of M as fixed points); furthermore compactness and convexity of M are essential, as simple counterexamples show.

Strategy to Satisfy the Requirements of Schauder's Theorem

In our specific case, $X = \vec{C}^0[t_0, t_1]$, the space of continuous vector functions on the interval, equipped with the usual maximum norm, and A is the integral operator in eq. (5). From continuity of $\vec{F}$, it follows easily that A is continous on X. The process of our application of Schauder's theorem now has three major steps:

1. Determine a sufficiently large family Y of subsets of X from which to draw candidates for the set M. To satisfy the requirements of Schauder's theorem, the sets in Y have to be compact and convex; and to fit within our computational framework, it should be possible to contain them in suitable Taylor models.

2. Using the differential algebraic structure on Taylor models, construct an initial set $M_0 \in Y$ that satisfies the inclusion property $A(M_0) \subset M_0$. Once this set has been determined, all requirements of the fixed point theorem are satisfied, and the existence of a solution in M_0 and hence within a Taylor model has been established.

3. Finally, the set M_0 is iteratively reduced in size in order to obtain a bound that is as sharp as possible. For $i = 1, 2, 3, ...$ we construct the sequence $M_i = A(M_{i-1})$. We have the chain $M_1 \supset M_2 \supset ...$, and we continue to iterate until no significant further reduction in size is possible.

Schauder Candidate Sets

For the first step, it is necessary to establish a family of sets Y from which to draw candidates for M_0. We define Y in the following way. Let $(\vec{P} + \vec{I})$ be a Taylor model depending on time as well as the initial condition $\vec{r}_0$. Then we define the associated set $M_{\vec{P}+\vec{I}}$ as follows:

$$M_{\vec{P}+\vec{I}} \subset \vec{C}^0[t_0, t_1]; \text{ and for } \vec{r} \in M_{\vec{P}+\vec{I}}:$$
$$\vec{r}(t_0) = \vec{r}_0$$
$$\vec{r}(t) \in \vec{P} + \vec{I} \; \forall t \in [t_0, t_1] \; \forall \vec{r}_0$$
$$|\vec{r}(t') - \vec{r}(t'')| \leq k|t' - t''| \; \forall t', t'' \in [t_0, t_1] \; \forall \vec{r}_0,$$

where in the last condition, k is a bound for $|\vec{F}|$ on the bounded set $M_{\vec{P}+\vec{I}}$, which exists because $\vec{F}$ is continuous; obviously k depends on $\vec{P}$ and $\vec{I}$. The last condition means that all $\vec{r} \in M_{\vec{P}+\vec{I}}$ are uniformly Lipschitz with constant k. Define the family of candidate sets Y as $Y = \bigcup_{\vec{P}+\vec{I}} M_{\vec{P}+\vec{I}}$

Convexity, Compactness and Invariance of Schauder Candidate Sets

Let $M \subset Y$ be a Schauder Candidate Set. Then M is convex because

$$\vec{x}_1, \vec{x}_2 \in M \Rightarrow$$
$$\alpha \vec{x}_1 + (1-\alpha)\vec{x}_2 \in M \; \forall \alpha \in [0, 1],$$

as any such linear combination of two k-Lipschitz functions is k-Lipschitz, is in the same Taylor models as $\vec{x}_1$ and $\vec{x}_2$, and assumes the value $\vec{r}_0$ at t_0.

Furthermore, M is compact, i.e. any sequence in M has a clusterpoint in M. To see this, let $(\vec{x}_n)$ be a sequence of functions in M. Then all $\vec{x}_n$ are k-Lipschitz and hence uniformly equicontinuous; since they are in the same Taylor model, they are uniformly bounded. Thus according to the Ascoli-Arzela Theorem, $(\vec{x}_n)$ has a uniformly convergent subsequence. Let $\vec{x}^*$ be the limit of this subsequence. Since the $\vec{x}_n$ are continous, so is $\vec{x}^*$, and we obviously have $\vec{x}^*(t_0) = \vec{r}_0$. Since the elements of the subsequence converging to $\vec{x}^*$ are k-uniformly Lipschitz, so is $\vec{x}^*$ itself, as a simple indirect proof reveals. Similarly, since the subsequence converging to $\vec{x}^*$ is in $\vec{P} + \vec{I}$, so is $\vec{x}^*$.

Finally, the images under A of the functions in $M_{\vec{P}+\vec{I}}$ are continuous because they are integrals. They go through $\vec{r}_0$ at t_0, and are k-Lipschitz because $\vec{F}$ is bounded by k. Hence all requirement of Schauder's fixed point theorem are met if we can find a Taylor model $\vec{P} + \vec{I}$ such that all continuous functions in $\vec{P} + \vec{I}$ are mapped into $\vec{P} + \vec{I}$; or in other words, if

$$A(\vec{P} + \vec{I}) \subset \vec{P} + \vec{I}. \tag{6}$$

Because if this condition is satisfied, then indeed we also have

$$A(M_{\vec{P}+\vec{I}}) \subset M_{\vec{P}+\vec{I}}.$$

But condition (6) can be verified computationally in a rigorous fashion using the differential algebraic representation of the operator A on the set of Taylor models!

Satisfying the Schauder Inclusion Requirement with Differential Algebraic Methods

For practical purposes it is of course in addition desirable to have I small. For this purpose it turns out to be important to determine a starting candidate that is on the one hand sufficiently small in width, but on the other hand shaped in such a way as to contain the true solution. This thought leads to attempt sets M^* of the form

$$M^* = M_{\mathcal{M}_n(\vec{r},t)+\vec{I}^*}, \tag{7}$$

where $\mathcal{M}_n(\vec{r}, t)$ is n-th order Taylor expansion of the solution. If n is large enough, we may expect that the true solution of the ODE is sufficiently close to the $n-$th order expansion, and hence that it may be possible to choose $\vec{I}^*$ rather small.

This approach requires the knowledge of the solution $\mathcal{M}_n(\vec{r}, t)$, and contrary to the usual situation in which we are only interested in $\mathcal{M}_n(\vec{r}, t)$ at the final value of t, here the explicit dependence on t is required. This quantity can be obtained by iterating eq. (5) within the DA of Truncated Power Series. To this end, one chooses an initial function

$$\mathcal{M}_n^{(0)}(\vec{r}, t) = \mathcal{I},$$

where $\mathcal{I}$ is the identity function, and then iteratively sets

$$\mathcal{M}_n^{(k+1)} =_n A(\mathcal{M}_n^{(k)}).$$

This process converges to the exact DA result $\mathcal{M}_n$ in $(n+1)$ steps.

Next, we try to find $\vec{I}^*$ such that in fact $A(\mathcal{M}_n(\vec{r}, t) + \vec{I}^*) \subset \mathcal{M}_n(\vec{r}, t) + \vec{I}^*$, the inclusion property necessary for Schauder's theorem. The suitable choice of $\vec{I}^*$ requires a little experimenting, it is however greatly simplified by the observation that it is necessary that computationally,

$$\vec{I}^* \supset \vec{I}_0 = A(\mathcal{M}_n(\vec{r}, t) + [0, 0]).$$

We may expect that $\vec{I}_0$ is a good benchmark for the size of intervals that is to be encountered; and so we iteratively try the sequence

$$\vec{I}^{(k)} = 2^k \cdot \vec{I}_0,$$

until a computational inclusion can be found, which means that we have established

$$A(\mathcal{M}_n(\vec{r}, t) + \vec{I}^{(k)}) \subset \mathcal{M}_n(\vec{r}, t) + \vec{I}^{(k)}. \tag{8}$$

Once this computational inclusion has been determined, a solution of the ODE is proven to exist within the Taylor model $\mathcal{M}_n(\vec{r}, t) + \vec{I}^{(k)}$, satisfying our demand. On the other hand, should it not be possible to find a computational inclusion, then with the current choice of the order n, it is not possible to prove the existence of a solution over the current size of domain intervals; in this case it is necessary to increase the order n, or to decrease the time step.

Iterative Refinement of the Inclusion

For practical purposes it is useful to note that the sharpness of this solution can be improved. Denoting $\vec{I}_1 = \vec{I}^{(k)}$, we iteratively define a sequence of Taylor models

$$\mathcal{M}_n(\vec{r},t) + \vec{I}_k = A(\mathcal{M}_n(\vec{r},t) + \vec{I}_{k-1}). \qquad (9)$$

We then must have $\vec{I}_k \subset \vec{I}_{k-1}$ for all $k = 1, 2, ...$ To see this, we observe that by definition of $\vec{I}_1$, this is the case for $k = 1$, and then we infer inductively

$$\mathcal{M}_n(\vec{r},t) + \vec{I}_k \subset \mathcal{M}_n(\vec{r},t) + \vec{I}_{k-1} \Rightarrow$$
$$A(\mathcal{M}_n(\vec{r},t) + \vec{I}_k) \subset A(\mathcal{M}_n(\vec{r},t) + \vec{I}_{k-1}) \Rightarrow$$
$$\mathcal{M}_n(\vec{r},t) + \vec{I}_{k+1} \subset \mathcal{M}_n(\vec{r},t) + \vec{I}_k.$$

But furthermore, the fixed point function $\vec{r}$ must actually be contained in each of the elements of the sequence of Taylor models $\mathcal{M}_n(\vec{r},t) + \vec{I}_k$. In fact, again by definition it is contained in $\mathcal{M}_n(\vec{r},t) + \vec{I}_1$, and by induction we see

$$\vec{r} \in \mathcal{M}_n(\vec{r},t) + \vec{I}_k \Rightarrow$$
$$A(\vec{r}) \in A(\mathcal{M}_n(\vec{r},t) + \vec{I}_k) \Rightarrow$$
$$\vec{r} \in \mathcal{M}_n(\vec{r},t) + \vec{I}_{k+1}$$

So this provides a mechanism to iteratively refine the inclusion until no further worthwhile decrease in size can be obtained.

Example

To show the use of the method in practice, we provide a first example of the method. We analyze the motion of a charged particle in a magnet with constant magnetic field over an extended phase space. Since the motion in the dipole can be solved analytically based on simple geometrical arguments related to intersections of circles and straight lines, this represents a useful check of the practical validity of the remainder bounds. For our example, we chose a magnet with a deflection radius $R = 1m$. The integration was carried out over a deflection angle of 36 degrees with a fixed step size of 4 degrees. The initial conditions are within the domain intervals

$$[-.02, .02] \times [-.02, .02] \times [-.02, .02] \times [-.02, .02],$$

and the Taylor polynomial describing the dependence of the four final coordinate values on the four initial coordinate values was determined. The order in time and initial conditions was chosen to be 12, and the step size was estimated so as to ascertain an overall accuracy below 10^{-9}; since no automatic step size control was utilized, the estimate proved conservative and the actual resulting remainder bounds were somewhat smaller:

$$[-0.4496880372277553E - 09, +0.3888593417126594E - 09]$$
$$[-0.1301070602141642E - 09, +0.1337099965985420E - 09]$$
$$[-0.3417079805637740E - 10, +0.3417079805637740E - 10]$$
$$[-0.0000000000000000E + 00, +0.0000000000000000E + 00].$$

The resulting Taylor polynomials describing the dependence of final on initial coordinates were compared with those obtained by the code COSY INFINITY[11],[9] and agreement was found. Furthermore, a program was written that solves the geometry for individual rays, and its results were compared for a large collection of rays with the results of the flow calculated by the verified integrator. For all rays studied, the difference between the final coordinates determined geometrically and those predicted by the twelfth order Taylor polynomial were within the calculated remainder bounds.

Acknowledgments

I would like to thank Kyoko Makino for many contributions to the Taylor model method, and in particular the implementation in COSY INFINITY and the calculations in this paper. This work was supported in part by the US Department of Energy and an Alfred P. Sloan Fellowship.

References

[1] M. Berz. The new method of TPSA algebra for the description of beam dynamics to high orders. Technical Report AT-6:ATN-86-16, Los Alamos National Laboratory, 1986.

[2] M. Berz. The method of power series tracking for the mathematical description of beam dynamics. *Nuclear Instruments and Methods*, A258:431, 1987.

[3] M. Berz. Differential algebraic description of beam dynamics to very high orders. *Particle Accelerators*, 24:109, 1989.

[4] K. L. Brown. The ion optical program TRANSPORT. Technical Report 91, SLAC, 1979.

[5] H. Wollnik, J. Brezina, and M. Berz. GIOS-BEAMTRACE, a computer code for the design of ion optical systems including linear or nonlinear space charge. *Nuclear Instruments and Methods*, A258:408, 1987.

[6] T. Matsuo and H. Matsuda. Computer program TRIO for third order calculations of ion trajectories. *Mass Spectrometry*, 24, 1976.

[7] A. J. Dragt, L. M. Healy, F. Neri, and R. Ryne. MARYLIE 3.0 - a program for nonlinear analysis of accelerators and beamlines. *IEEE Transactions on Nuclear Science*, NS-3,5:2311, 1985.

[8] M. Berz, H. C. Hofmann, and H. Wollnik. COSY 5.0, the fifth order code for corpuscular optical systems. *Nuclear Instruments and Methods*, A258:402, 1987.

[9] K. Makino and M. Berz. COSY INFINITY Version 7. In *Fourth Computational Accelerator Physics Conference*. AIP Conference Proceedings, 1996.

[10] M. Berz, G. Hoffstätter, W. Wan, K. Shamseddine, and K. Makino. COSY INFINITY and its applications to nonlinear dynamics. *in: Computational Differentiation: Techniques, Applications, and Tools, M. Berz, C. Bischof, G. Corliss, A. Griewank (Eds.)*, SIAM, 1996.

[11] M. Berz. COSY INFINITY Version 7 reference manual. Technical Report MSUCL-977, National Superconducting Cyclotron Laboratory, Michigan State University, East Lansing, MI 48824, 1996. see also http://www.beamtheory.nscl.msu.edu/cosy.

[12] M. Berz. COSY INFINITY Version 6. In *M. Berz, S. Martin and K. Ziegler (Eds.), Proc. Nonlinear Effects in Accelerators*, page 125. IOP Publishing, 1992.

[13] L. Michelotti. MXYZTPLK: A practical, user friendly c++ implementation of differential algebra. Technical report, Fermilab, 1990.

[14] W.G. Davis, S. R. Douglas, G. D. Pusch, and G. E. Lee-Whiting. The Chalk River differential algebra code DACYC and the role of differential and Lie algebras in understanding the orbit dynamics in cyclotrons. In *M. Berz, S. Martin and K. Ziegler (Eds.), Proceedings Workshop on Nonlinear Effects in Accelerators*. IOP Publishing, 1993.

[15] J. van Zeijts and F. Neri. The arbitrary order design code TLIE 1.0. In *M. Berz, S. Martin and K. Ziegler (Eds.), Proceedings Workshop on Nonlinear Effects in Accelerators*. IOP Publishing, 1993.

[16] J. van Zeijts. New features in the design code TLIE. In *Third Computational Accelerator Physics Conference*, page 285. AIP Conference Proceedings 297, 1993.

[17] Y. Yan and C.-Y. Yan. ZLIB, a numerical library for differential algebra. Technical Report 300, SSCL, 1990.

[18] Y. Yan. ZLIB and related programs for beam dynamics studies. In *Third Computational Accelerator Physics Conference*, page 279. AIP Conference Proceedings 297, 1993.

[19] F. C. Iselin. The CLASSIC project. In *Fourth Computational Accelerator Physics Conference*. AIP Conference Proceedings, 1996.

[20] M. Berz. *Truncated Power Series Techniques, Entry in 'Handbook of Accelerator Physics and Engineering', M. Tigner and A. Chao (Eds.)*. World Scientific, New York, in preparation, 1997.

[21] M. Berz. *High-Order Computation and Normal Form Analysis of Repetitive Systems*, in: M. Month (Ed), *Physics of Particle Accelerators*, volume AIP 249, page 456. American Institute of Physics, 1991.

[22] M. Berz. Differential algebraic description of beam dynamics to very high orders. Technical Report SSC-152, SSC Central Design Group, Berkeley, CA, 1988.

[23] M. Berz. Symplectic tracking in circular accelerators with high order maps. In *Nonlinear Problems in Future Particle Accelerators*, page 288. World Scientific, 1991.

[24] A. J. Dragt and J. M. Finn. Lie series and invariant functions for analytic symplectic maps. *Journal of Mathematical Physics*, 17:2215, 1976.

[25] A. J. Dragt and J. M. Finn. Normal form for mirror machine Hamiltonians. *Journal of Mathematical Physics*, 20(12):2649, 1979.

[26] A. Bazzani. Normal forms for symplectic maps on R^{2n}. *Celestial Mechanics*, 42:107–128, 1988.

[27] E. Forest, M. Berz, and J. Irwin. Normal form methods for complicated periodic systems: A complete solution using Differential algebra and Lie operators. *Particle Accelerators*, 24:91, 1989.

[28] M. Berz. Calculus and numerics on Levi-Civita fields. *in: Computational Differentiation: Techniques, Applications, and Tools*, SIAM, 1996.

[29] M. Berz. Analysis auf einer nichtarchimedischen Erweiterung der reellen Zahlen. Report (in German) MSUCL-753, Department of Physics, Michigan State University, 1990.

[30] M. Berz. Analysis on a nonarchimedean extension of the real numbers. Lecture Notes, 1992 and 1995 Mathematics Summer Graduate Schools of the German National Merit Foundation. MSUCL-933, Department of Physics, Michigan State University, 1994.

[31] K. Shamseddine and M. Berz. Exception handling in derivative computation with nonarchimedean calculus. *in: Computational Differentiation: Techniques, Applications, and Tools*, M. Berz, C. Bischof, G. Corliss, A. Griewank (Eds.), SIAM, 1996.

[32] K. Shamseddine and M. Berz. Nonarchimedean structures as differentiation tools. *1997 Beirut International Conference on Numerical Analysis*, 1997.

[33] J. F. Ritt. *Integraton in Finite Terms - Liouville's Theory of Elementary Methods*. Columbia University Press, New York, 1948.

[34] J. F. Ritt. *Differential Equations from the Algebraic Viewpoint*. American Mathematical Society, Washington, D.C., 1932.

[35] E. R. Kolchin. *Differential Algebra and Algebraic Groups*. Academic Press, New York, 1973.

[36] R. H. Risch. The problem of integration in finite terms. *Transactions of the American Mathematical Society*, 139:167–189, 1969.

[37] R. H. Risch. The solution of the problem of integration in finite terms. *Bulletin of the American Mathematical Society*, 76:605–608, 1970.

[38] R. H. Risch. Algebraic properties of the elementary functions of analysis. *American Journal of Mathematics*, 101 (4):743–759, 1979.

[39] W. F. Feehery and P. I. Barton. A differentiation-based approach to dynamic simulation and optimization with high-index differential-algebraic equations. *in: Computational Differentiation: Techniques, Applications, and Tools, M. Berz, C. Bischof, G. Corliss, A. Griewank (Eds.)*, SIAM, 1996.

[40] M. Berz, C. Bischof, A. Griewank, and G. Corliss (Eds.). *Computational Differentiation: Techniques, Applications, and Tools*. SIAM, Philadelphia, 1996.

[41] A. Griewank and G. F. Corliss (Eds.). *Automatic Differentiation of Algorithms*. SIAM, Philadelphia, 1991.

[42] M. Berz. *Computational Differentiation, Entry in 'Encyclopedia of Computer Science and Technology'*. Marcel Dekker, New York, in preparation, 1997.

[43] J. Abate, C. Bischof, L. Roh, and A. Carle. Algorithms and design for a second-order automatic differentiation module. In *Proceedings, ISSAC 97, Maui*, 1997. also available at ftp://info.mcs.anl.gov/pub/tech_reports/reports/P636.ps.Z.

[44] C. Bischof and Fred Dilley. A compilation of automatic differentiation tools. http://www.mcs.anl.gov/autodiff/AD_Tools/index.html. www html page.

[45] M. Berz. Forward algorithms for high orders and many variables. *Automatic Differentiation of Algorithms: Theory, Implementation and Application*, SIAM, 1991.

[46] M. Berz. The Differential algebra FORTRAN precompiler DAFOR. Technical Report AT-3:TN-87-32, Los Alamos National Laboratory, 1987.

[47] M. Berz. Differential algebra precompiler version 3 reference manual. Technical Report MSUCL-755, Michigan State University, East Lansing, MI 48824, 1990.

[48] M. Berz and G. Hoffstätter. Computation and application of Taylor polynomials with interval remainder bounds. *Interval Computations, in print*, 1997.

[49] K. Makino and M. Berz. Remainder differential algebras and their applications. *in: Computational Differentiation: Techniques, Applications, and Tools, M. Berz, C. Bischof, G. Corliss, A. Griewank (Eds.)*, SIAM, 1996.

[50] M. Berz and G. Hoffstätter. Exact bounds of the long term stability of weakly nonlinear systems applied to the design of large storage rings. *Interval Computations*, 2:68–89, 1994.

[51] J. B. Conway. *Functional Analysis*. Springer, 1990.

[52] M. Berz. *Higher Order Derivatives and Taylor Models, Entry in 'Encyclopedia of Optimization'*. Kluwer, Boston, 1997.

Numerical Evaluation of Long-term Stability

M. Giovannozzi[‡], W. Scandale[†] and E. Todesco[‡]

[†] *CERN, SL Division, CH 1211 Geneva Switzerland*
[‡] *University of Bologna, Department of Physics, Via Irnerio 46, 40126, Bologna Italy*

Abstract. The problem of predicting long-term particle loss in 4D betatronic motion is considered. A phenomenological scenario is derived through numerical tools based on tracking and frequency analysis. A three-parameter formula to interpolate the dynamic aperture versus the number of turns is proposed. The agreement with tracking data is excellent, and the extrapolation for very high number of turns agrees with the onset of chaos evaluated through the Lyapunov method.

INTRODUCTION

Modern hadron colliders based on superconducting magnets suffer from the unavoidable effect of field-shape distortions, particularly harmful during the injection plateau. This critical period can last a considerable number of turns making difficult to evaluate the single-particle stability with computer tracking simulations. For instance, in the case of the CERN Large Hadron Collider [1], the injection process will last of approximately 10^7 turns. On the other hand, numerical simulations based on symplectic tracking can hardly reach $10^5 - 10^6$ turns. In addition, a dense sampling of the phase space is crucial to obtain significant results from numerical tracking. Three main approaches have been proposed in the past to speed-up the investigations on beam stability: the determination of the onset of chaotic behaviour using the maximal Lyapunov exponent [2,3], the evaluation of the drift in the space of approximated invariants carefully evaluated through numerical methods [4], and the visualization of the dynamic aperture reduction with increasing number of turns through survival plots [6,7].

In this paper we investigate extensively the last one using a simplified model of 4D betatronic motion where the coupling with longitudinal dynamics and the modulation of the linear frequencies are neglected. We propose some numerical tools and we derive a phenomenological scenario to interpret the results of our simulations. We recall a way to define the dynamic aperture [8] and we show how it can be interpolated using a three-parameter formula that can be justified in terms of the Nekhoroshev and KAM theorem. The interpolation fits very well with the numerical data and agrees with the prediction of the onset of chaos provided by the Lyapunov exponent. Additional studies to check the validity of this scenario in a more realistic accelerator model, describing 6D motion and including the ripple effect, are in progress [9,10].

MODELS

In this paper we restrict ourselves to analyse betatron motion neglecting the effect of coupling with the synchrotron motion. Therefore, the map that simulates the single-particle dynamics over one turn of the machine has a four dimensional phase space, and its linear part is the direct product of two rotations of frequencies ω_x and ω_y. A prototype of these models is the Hénon map that represents a linear lattice with a sextupole in the one-kick approximation:

$$\begin{pmatrix} x' \\ p'_x \\ y' \\ p'_y \end{pmatrix} = \begin{pmatrix} R(\omega_x) & 0 \\ 0 & R(\omega_y) \end{pmatrix} \begin{pmatrix} x \\ p_x + x^2 - y^2 \\ y \\ p_y - 2xy \end{pmatrix}. \quad (1)$$

We made several simulations over this model for different values of the linear frequencies. In this paper we will only present the results relative to this model. Simulations over a 4D model of the LHC have confirmed the results obtained for the Hénon map [3].

ANALYSIS OF THE PHASE SPACE

In this section we review some numerical tools based on frequency analysis [3,11–13] to determine the position and the width of resonances in the 4D phase space. We consider a grid of initial conditions on the plane (x, y), $p_x = p_y = 0$. The numerical method is based on two tools: the orbit of length N associated to each initial condition is evaluated through the iteration of the map, and then the main frequencies are computed using an interpolation of the FFT plus Hanning filter [13]. This methods provides, under some general conditions, a precision of the order of $1/N^4$ for regular trajectories (where the frequencies are well defined). For chaotic orbits the frequencies are not defined and the algorithm provides quantities that can considerably vary along the discrete time N, and that do not converge for $N \to \infty$.

For each initial condition, the simulations provide the number of turns n up to which the particle is stable and the two nonlinear frequencies (ν_x, ν_y). One can display this information through the following plots.

- Long-term plot: each initial condition (x, y) is plotted using a different marker according to the number of turns n at which particle loss occurs. One can visualize the shape of the dynamic aperture and how it shrinks when the length of the orbit N increases.

- Tune footprint (or image of the frequency map): the frequencies are plotted in the frequency space. This provides very relevant information about what part of the tune diagram is covered by the stable initial conditions. Moreover, depletion regions around resonance lines indicate that the resonance is excited and unstable.

- Network of resonances: only the initial conditions that are locked on resonances, i.e. whose frequencies satisfy

$$q\nu_1 + p\nu_2 = l + \epsilon \qquad q, p, l \in \mathbf{Z} \qquad (2)$$

are plotted in the coordinate plane (x, y). We used 2048 iterates and $\epsilon = 10^{-4}$. This plot directly displays the size of resonances, their position in phase space, and their relation with the dynamic aperture. Contrary to the tune footprint, this plot is not invariant under the selection of the initial phases. Nevertheless the plot is invariant under the linear dynamics and therefore a change in the initial phases does not significantly affect the obtain pattern, leading only to some deformations at high amplitudes. Indeed, one can produce similar plot in the space of the nonlinear invariants [14] by evaluating them through numerical methods. In this way one can rigorously measure the position of the resonances and their width in phase space. We will not use this method since we are interested only in a qualitative analysis of the phase space dynamics.

In Fig. 1 we show a very dense long-term plot for the Hénon map [see Eq. (1)] with linear frequencies $\nu_x = 0.168$, $\nu_y = 0.201$. A rectangular grid of 250×250 initial conditions is iterated for 10^8 turns. The different shadows of grey correspond to the particle loss number. It turns out that one has an inner region where all the particle are stable for at least 10^8 turns; such a region features no holes, at least at the resolution used in the scan of the initial conditions. Then, one finds a rather irregular but sharp border of instability: outside the border, one has a chaotic sea of initial conditions that are lost between 10^8 and $10^3 - 10^2$ turns; no structure is visible in this region, and neighbour initial conditions can be lost at number of turns n that may differ by one or two order of magnitudes. Finally, there is an outer region of fast particle loss ($10 - 10^2$ turns) where the dependence of the initial conditions on

n is much smoother, and some structures probably associated with hyperbolic manifolds are visible.

In Fig. 2 we show the network of resonance of the same model. One finds very large resonances that are stable; moreover, the mechanism of particle loss due to the diffusion along the resonant channels or due to resonance crossing does not seem to be very relevant. Indeed, the bulk of long-term losses occur in the wide chaotic band where no resonance structures are visible. This chaotic band is characterized by isolated points locked on very low order resonances,

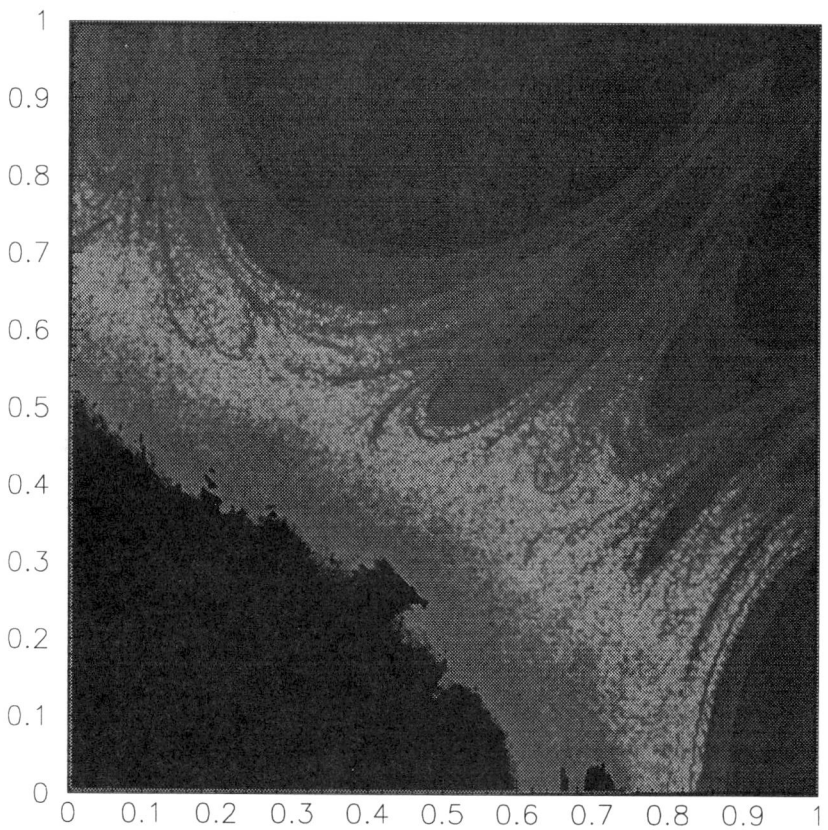

FIGURE 1. Long-term plot of the Hènon map at $\nu_x = 0.168$, $\nu_y = 0.201$: black dots represent initial conditions stable up to 10^8 turns; full circles represent unstable initial conditions (smaller circles correspond to shorter stability times).

that appear in the figure as a set of scattered dots. The same analysis has been carried out for other models, and also for a 4D model of the LHC [3], leading to the same qualitative results. In Figs. 3 and 4 we show the long-term plot and the resonance network for the Hénon map with linear frequencies $\nu_x = 0.201$, $\nu_y = 0.112$. Also in this case most long-term losses occur in the wide chaotic band localized at $x \in [0.25, 0.45]$, $y \in [0.0, 0.3]$ that can be recognized in the resonance network by the presence of isolated initial conditions locked on low-order resonances. Moreover, in this case one has the rather striking presence of a very large resonance that is stable for at least 10^7 turns. The size of this resonance cannot be worked out by analyzing the tune footprint, that only displays a strong phase locking around that resonance.

The conclusions of this qualitative analysis are the followings.

- There is a rather sharp border that separates stable from unstable initial conditions for the considered number of turns ($10^7 - 10^8$).

- Long-term particle losses mainly occur in wide chaotic bands where all the integrable structure has been wiped out.

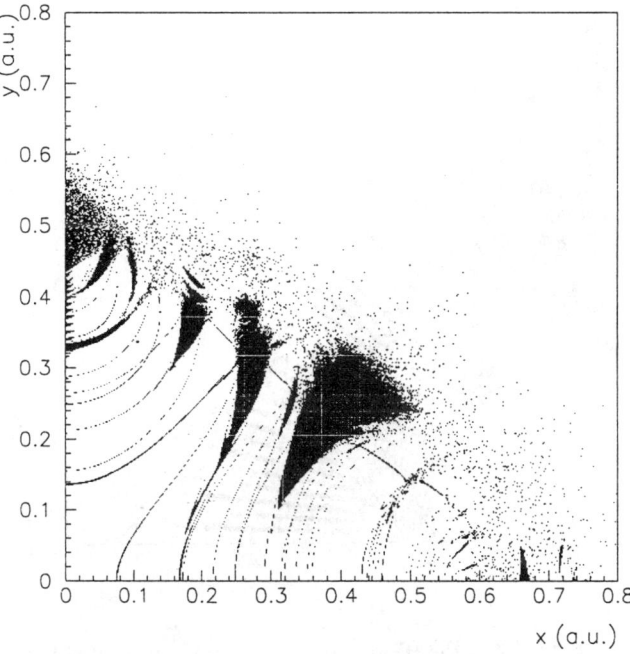

FIGURE 2. Network of resonances of the Hènon map at $\nu_x = 0.168$, $\nu_y = 0.201$: black dots represent initial conditions locked on resonances up to order 15.

- The mechanism of diffusion along the resonant channels and due to resonance crossing are rather weak.

DYNAMIC APERTURE AND ASSOCIATED ERRORS

In a previous work [8] we have proposed a definition of dynamic aperture as a function of the number of turns N as the first amplitude where particle loss occurs before N turns, averaged over the phase space. Particles are started along a 2D polar grid in the coordinate space (x, y):

$$x = r \cos \theta \qquad y = r \sin \theta \qquad (3)$$

and the intial momenta p_x p_y are set to zero. Let $r(\theta; N)$ be the last stable initial condition along θ before the first loss at a turn number lower than N occurs. Then the dynamic aperture is defined as

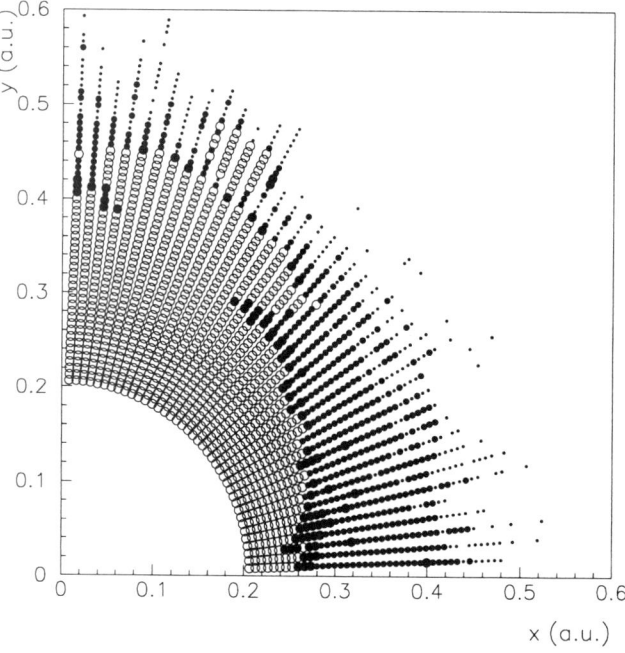

FIGURE 3. Long-term plot of the Hènon map at $\nu_x = 0.201$, $\nu_y = 0.112$: empty circles represent initial conditions stable up to 10^7 turns; full circles represent unstable initial conditions (smaller circles correspond to shorter stability times).

$$D = \left(\int_0^{\pi/2} [r(\theta; N)]^4 \sin 2\theta d\theta \right)^{1/4}. \tag{4}$$

With respect to the approach used in several long-term simulations (see for instance [6,7]), where the scan over only one variable is considered in order to speed up simulations, this definition provides a smoother dependence of D on N, thus allowing to derive interpolating formulas and to extrapolate them to predict long-term particle loss.

One of the crucial issues in the definition of the dynamic aperture is the determination of the error associated with the estimate of the dynamic aperture. When definition (4) is implemented in a computer code, one has to carry out two discretizations: one over the radial variable r and one over the angular variable θ. Let $\Delta r = (r_{max} - r_{min})/N_r$ and $\Delta \theta = \pi/2N_\theta$ be the step size in r and θ respectively. The total error can be obtained using gaussian sum in quadrature

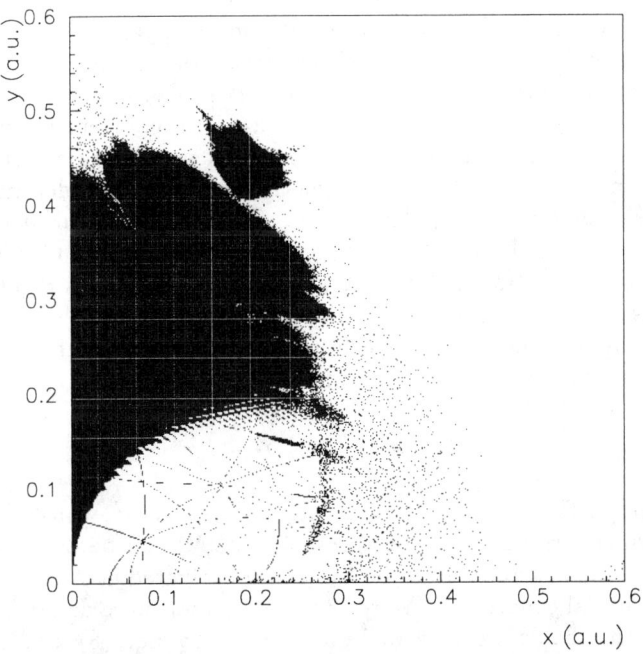

FIGURE 4. Network of resonances of the Hènon map at $\nu_x = 0.201$, $\nu_y = 0.112$: black dots represent initial conditions locked on resonances up to order 15.

$$\Delta D = \sqrt{\left(\frac{\partial D}{\partial r}\frac{\Delta r}{2}\right)^2 + \left(\frac{\partial D}{\partial \theta}\frac{\Delta \theta}{2}\right)^2}. \qquad (5)$$

An approximated formula for the error can be obtained by replacing the dynamic aperture definition with a simple average over θ

$$D = \frac{2}{\pi}\int_0^{\pi/2} r(\theta; N)d\theta \equiv <r(\theta; N)>. \qquad (6)$$

Using this formula the associated error reads

$$\Delta D = \sqrt{\frac{(\Delta r)^2}{4} + <|\frac{\partial r}{\partial \theta}|>^2 \frac{(\Delta \theta)^2}{4}} \qquad (7)$$

and therefore it turns out that the step in r must be equal to the step in θ times $<|\frac{\partial r}{\partial \theta}|>$ in order to have an optimization of the integration steps.

PREDICTION BASED ON LYAPUNOV EXPONENT

A method that has been used to select chaotic from regular orbits in nonlinear dynamical systems is based on the evaluation of the maximal Lyapunov exponent [2,3,15–17]. For a given initial condition, the Lyapunov $\lambda(N)$ is evaluated using an orbit of N turns. The theory states that if $\lim_{N\to\infty} \lambda(N) = 0$, the orbit is regular and therefore the particle is stable; on the other hand, if the limit is positive, then the particle is chaotic (i.e., there is sensitivity to initial conditions and exponential divergence of nearby trajectories), and therefore it can be lost. The Lyapunov method allows one to determine the border between chaotic and regular motion. On the other hand, it does not give information about how this limit is reached when N tends to infinity.

In Ref. [3] we have proposed an automatic method to select regular from chaotic orbits based on a threshold on the Lyapunov. Since for regular particles the distance between neighbour orbits linearly increases with the discrete time N, one can fix a threshold according to

$$\sigma_\lambda(N) = \frac{1}{N}\log A_\lambda N \qquad (8)$$

If $\lambda(N) > \sigma_\lambda(N)$, then the particle is assumed to be chaotic, whilst if $\lambda(N) < \sigma_\lambda(N)$ the particle is regular. The thresholds $\sigma_\lambda(N)$ can be determined by the analysis of the distribution of the Lyapunov evaluated at N turns for the chosen set of initial conditions. We refer to Ref. [3] for more details. It turns out that the thresholds are very well interpolated by Eq. (8), and that the constant A_λ seems to depend very weakly on the model. In fact, we found for all our simulations (Hénon, SPS, LHC 4D, LHC 6D, see Ref. [3]) the value $A_\lambda = 0.5$.

PREDICTION BASED ON EXTRAPOLATION

During the past years, extensive tracking has been carried out to determine the long-term stability of both existing and planned machines. A very effective way to display the tracking data is provided by survival plots [6,7], where the particle loss number n is plotted versus the initial amplitude A for a given lattice. The obtained pattern is in general rather irregular; in fact, due to the complicated structure of the chaotic region (see the previous sections), the dependence of n versus the amplitude is far from being regular. Therefore, an extrapolation of the amplitude for higher loss numbers is very hard to obtain. The definition of dynamic aperture given in the previous section allows one to make the survival plot considerably smoother, and therefore an interpolation becomes possible. In Fig. 5 we show $D(N)$ versus N for the same model of Fig. 1, carrying out simulations up to 10^8 turns. A very fine phase space scan (120 radial steps from 0.3 to 0.8 and 60 angular step) has been used in order to obtain a very small error (of the order of 1%).

In Ref. [3] we have proposed an "inverse log" behaviour to interpolate the dynamic aperture:

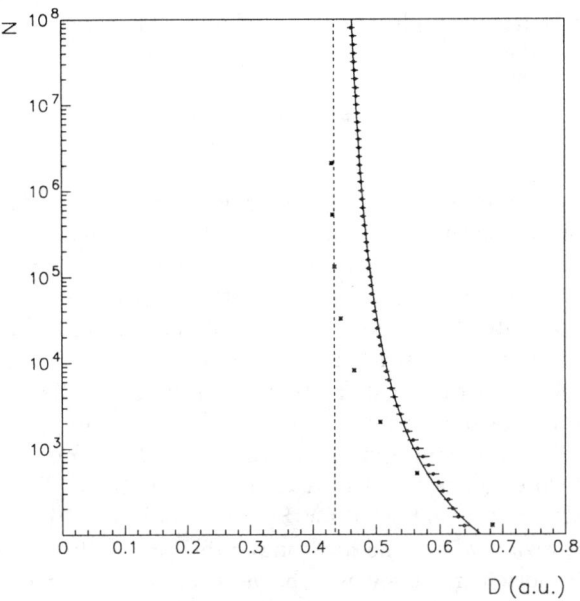

FIGURE 5. Dynamic aperture versus number of turns (dots) for the Hènon map at $\nu_x = 0.168$, $\nu_y = 0.201$; analytic interpolation (solid line) and extrapolation at infinity (dotted line). Prediction of the chaotic border according to the Lyapunov exponent (stars).

$$D(N) = D_\infty \left(1 + \frac{b}{\log N}\right). \tag{9}$$

An heuristic interpretation of this formula has been proposed [18] according to two main theorems of dynamical systems, namely KAM [19] and Nekhoroshev [20–23] theorems. One assumes that the phase space is divided into two regimes:

- An inner region where almost all the phase space is foliated into KAM tori, except a very small fraction where the Arnold diffusion can take place over the resonance web. This region appears in simulations as a "full" domain of initial conditions stable for extremely high number of turns.

- An outer region where almost all the foliation of phase space in KAM tori has been destroyed, and only a wide chaotic sea is left. Since we are close to the last KAM tori, we assume that in this region the particles escape to infinity with the rate provided by the Nekhoroshev estimate:

$$N(r) = N_0 \exp\left(\frac{r_*}{r}\right)^{1/\kappa}, \tag{10}$$

where $N(r)$ is the number of turns that are estimated to be stable for particles with initial amplitude smaller than r. The inversion of the above formula provides

$$r(N) = \frac{r_*}{\log^\kappa(N/N_0)} \tag{11}$$

In order to check out this scenario, we have carried out the following simulations over the model used in Fig. 1: we choose initial conditions along $x = y = A$ and $p_x = p_y = 0$, and for each amplitude A we start a dense cloud of initial conditions around A (1000 particles with a neighbourhood size of 10^{-4}). We compute the fraction $S(A; N)$ of the particles that are stable for at least N turns: assuming that all the chaotic particles are unstable, the $\lim_{N\to\infty} S(A; N)$ is the local fraction $S(A)$ of the phase space foliated into invariant tori as a function of the amplitude. If the proposed scenario is valid, one should find a good approximation of the theta function, i.e. $S(A)$ should be very close to one for $A < A_\infty$, and then it should fall abruptly to zero. The results shown in Fig. 6 for $S(A; 10^7)$ fully confirm this scenario: the phase space region where one has comparable probability of finding both KAM tori and chaotic regions seems to be very small. Moreover, for $A < A_\infty$, $S(A; 10^7)$ is exactly equal to one, i.e. for each A all the 10^3 particles were found to be stable. This confirms that the Arnold diffusion for this kind of models is extremely weak and therefore is not an important mechanism for the determination of the dynamic aperture.

Adding the information obtained from the KAM theorem (the existence of a positive D_∞) and from the Nekhoroshev theorem (the inverse log decaying of the dynamic aperture), one obtains the following equation

$$D(N) = D_\infty \left(1 + \frac{b}{\log^\kappa(N/N_0)}\right) \quad (12)$$

that reduces to Eq. 9 for $N_0 = \kappa = 1$. We tried to interpolate the data shown in Fig. 5 with this formula using three free parameters D_∞, b and the exponent κ. We fixed N_0 to one by using the heuristic argument that $D(1) = \infty$. In order to find a solution, we made a scan over κ and for each κ we evaluated D_∞, and b by solving a linear system. Then we computed the χ^2 function, i.e.

$$\chi^2 = \frac{1}{J-3} \sum_{j=1}^{J} \frac{(D(N_j) - \hat{D}(N_j))^2}{\sigma_i} \quad (13)$$

where the interpolated dynamic aperture $\hat{D}(N_i)$ according to Eq. (9) is evaluated at the turn number N_i, and $\sqrt{\sigma_i}$ is the error estimated through Eq. (7). It turns out that

- It is rather difficult to determine the exponent with a high precision. For instance, if we consider all the exponents that provide a χ smaller than 0.7, that corresponds in our case of a confidence level of 95%, we obtain $\kappa \in [0.9, 2]$.

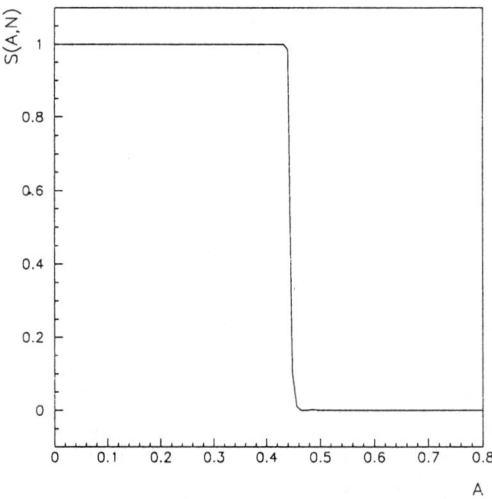

FIGURE 6. Fraction of the particles stable for 10^7 turns around the initial condition $x = y = A$, $p_x = p_y = 0$ for the Hènon map at $\nu_x = 0.168$, $\nu_y = 0.201$.

- The optimal exponent for our case turns out to be around 1.5. The interpolation is shown in Fig. 5 as a solid line, and agrees very well with tracking data. The theory of Nekhoroshev for mappings [22,23] provides an estimate for the exponent κ that is equal to the number of degrees of freedom plus one, i.e. in our case $\kappa = 3$. This value of κ does not agree with our simulations. Indeed, a refined version of the theorem [24] leads to the improved estimate $\kappa = (1 + d)/2$, i.e. in our case $\kappa = 1.5$. This is in agreement with our simulations; additional checks for higher dimensions would be highly desirable in order to cross-check the optimal estimate of the exponent with the validity of our scenario.

We have also computed the estimate of the chaotic border through the Lyapunov exponent, using the same type of definition for the dynamic aperture, where now $r(\theta)$ the amplitude of the particle immediately before the first particle along θ whose Lyapunov is greater than the threshold. In Fig. 5 we show, together with the tracking data and with the interpolation, the guess provided by the Lyapunov exponent at increasing number of turns. It turns out that the Lyapunov guess of the chaotic border seems to saturate rather rapidly (when

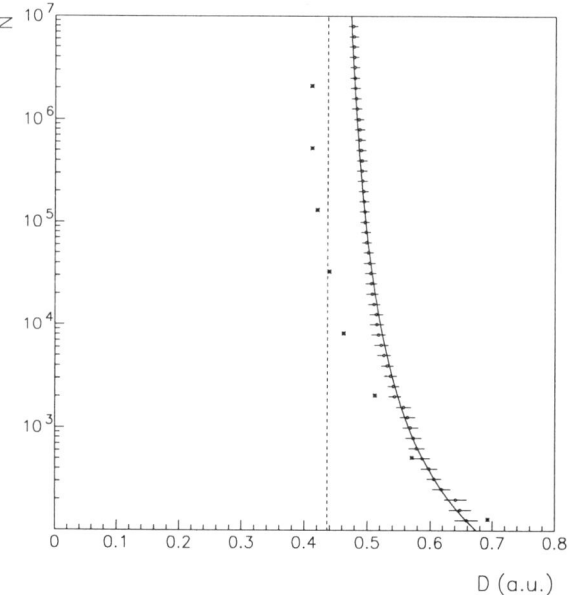

FIGURE 7. Dynamic aperture versus number of turns (dots) for the Hènon map at $\nu_x = 0.201$, $\nu_y = 0.168$; analytic interpolation (solid line) and extrapolation at infinity (dotted line). Prediction of the chaotic border according to the Lyapunov exponent (stars).

compared to tracking) to the dynamic aperture value extrapolated to infinity according to our formula. This result confirms the scenario illustrated in the previous section. The same simulations have been carried out for other two Hénon maps with different linear frequencies. The dynamic aperture data, the interpolation and the Lyapunov prediction are shown in Fig. 7 and 8. The results fully agree with the scenario described for the previous case.

CONCLUSION

We have analysed simplified models of the 4D betatronic motion to derive a phenomenological description of the mechanisms of instability and to build numerical methods to predict long-term particle loss. Long-term tracking of very dense sets of initial conditions shows that particle loss mainly occurs in macroscopic chaotic bands. Other mechanism such as Arnold diffusion or diffusion along resonances are shown to be rather weak. We also show that the dynamic aperture, computed as an average over the ratio of emittances, is very well interpolated by an empirical law that can be justified in terms of

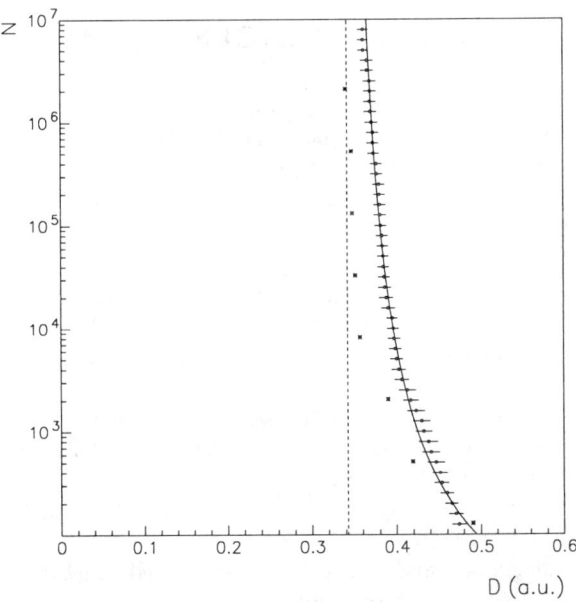

FIGURE 8. Dynamic aperture versus number of turns (dots) for the Hènon map at $\nu_x = 0.201$, $\nu_y = 0.112$; analytic interpolation (solid line) and extrapolation at infinity (dotted line). Prediction of the chaotic border according to the Lyapunov exponent (stars).

KAM and Nekhoroshev theorem. The determination of the exponent in the inverse log formula is affected by a rather large error, but the extrapolation is reliable. Moreover, the extrapolation for infinite number of turns agrees well with the guess provided by the Lyapunov method. Even though relevant studies for the 6D case has been already carried out [2,3,9], we believe that additional analysis should be done, including also the ripple effect, in order to see how the presence of an additional degree of freedom modifies the proposed scenario and the predictivity of the numerical tools.

ACKNOWLEDGEMENTS

We wish to thank Prof. Turchetti for very relevant contributions to the analysis and to the interpretation of the tracking data. Special thanks to S. Bongini, M. Böge, J. Ellison, J. Irwin, F. Schmidt and B. Warnock, for constructive discussions. E. Todesco would like to thank Z. Parsa and G. Guignard for the invitation to the program, the ITP staff for the very effective organization of the workshop, Anu Venugopalan, Vera and Raymond Holgate, and the management of "L'Agave" for the warm welcome to S. Barbara.

REFERENCES

1. The LHC Study Group, *CERN-LHC* **95-05** (1995).
2. Schmidt, F., Willeke, F., and Zimmermann, F. , *Part. Accel.* **35** 249 (1991).
3. Giovannozzi, M., Scandale, W., and Todesco, E., *Part. Accel.*, in press (1997).
4. Warnock, R. L., and Ruth, R. D., *Physica D* **56** 188 (1992).
5. Chao, A., *AIP Conf. Proc.* **230** 203 (1990).
6. Yan, Y. *SSC* **500** (1991).
7. Galluccio, F., and Schmidt, F., *Third European Particle Accelerator Conference* Gif sur Yvette: Edition Frontiéres (1993) pp. 640–42.
8. Todesco, E., Giovannozzi M., *Phys. Rev. E* **53** 4067 (1996).
9. Böge, M., and Schmidt, F., these proceedings.
10. Giovannozzi, M., Scandale, W., and Todesco, E., in preparation.
11. Laskar, J., Froeschlé, C., and Celletti, A., *Physica D* **56** 253 (1992).
12. Laskar, J., *Physica D* **67** 257 (1992).
13. Bartolini, R., Bazzani, A., Giovannozzi, M., Scandale, W., and Todesco, E., *Part. Accel.* **52** 147 (1996).
14. Bazzani, A., Bongini, S., and Turchetti, G., submitted to *Phys. Rev. E*.
15. Hénon, M., and Heiles, C., *Astr. J.* **69-1** 73 (1964).
16. Froeschlé, C., *Astron. & Astrophys.* **16** 172 (1972).
17. Benettin, G., Galgani, L., Giorgilli, A., and Strelcyn, J. M., *Meccanica* **15-1** 21 (1980).
18. Turchetti, G., private communication.

19. Siegel, C. L., and Moser, J., *Lectures in celestial mechanics*, Berlin: Springer Verlag, 1971.
20. Nekhoroshev, N., *Russ. Math. Surv.* **32** 1 (1977).
21. Benettin, G., Galgani, L., Giorgilli, A., Servizi, G., and Turchetti, G., *Phys. Lett.* **95A** 11 (1983).
22. Turchetti, G., *Number theory and physics* Berlin–Heidelberg: Springer Verlag 223 (1990).
23. Bazzani, A., Marmi S., and Turchetti, G., *Cel. Mech.* **47** 333 (1990).
24. Bazzani, A., and Turchetti, G., *Chaotic Dynamics: Theory and Practice* New York: Plenum Press 59 (1991).

Convergence of a Fourier-Spline Representation for the Full-turn Map Generator

Robert L. Warnock* and James A. Ellison[†]

*Stanford Linear Accelerator Center, Stanford, California, 94309, and [†]Department of Mathematics and Statistics, University of New Mexico, Albuquerque, New Mexico, 87131

Abstract. Single-turn data from a symplectic tracking code can be used to construct a canonical generator for a full-turn symplectic map. This construction has been carried out numerically in canonical polar coordinates, the generator being obtained as a Fourier series in angle coordinates with coefficients that are spline functions of action coordinates. Here we provide a mathematical basis for the procedure, finding sufficient conditions for the existence of the generator and convergence of the Fourier-spline expansion. The analysis gives insight concerning analytic properties of the generator, showing that in general there are branch points as a function of angle and inverse square root singularities at the origin as a function of action.

INTRODUCTION

Fast symplectic mapping is a powerful tool for study of long-term stability in accelerators, especially in large hadron storage rings such as the LHC (1),(2). Here we are concerned with a representation of the full-turn map in terms of a canonical mixed-variable generator, which can be constructed using many single-turn data from a symplectic tracking code (3). In numerical work to date, the generator has been expanded in a Fourier series in angle variables, with coefficients given as spline functions of action variables. We wish to find conditions so that this expansion converges (in the limit of infinitely many Fourier modes and spline interpolation points) to the exact generator of the full-turn evolution defined by the tracking code. We adopt canonical polar cooordinates (I, Φ), where I and Φ are n-component action and angle vectors, respectively. These are usually action-angle coordinates of an underlying linear system, but need not be such. The full-turn map $M : (I, \Phi) \mapsto (I', \Phi')$ as defined by the tracking code is denoted as follows:

$$I' = I + R(I, \Phi) , \tag{1}$$
$$\Phi' = \Phi + \Theta(I, \Phi) . \tag{2}$$

The existence of the inverse of the angular map (2) at fixed I is important in our analysis. We write it as

$$\Phi = \Phi' + F(I, \Phi') . \tag{3}$$

The function F is 2π-periodic in each component of Φ', as are R and Θ in each component of Φ. We assume that the tracking code is symplectic, so that the Jacobian matrix D satisfies

$$DJD^T = J , \tag{4}$$

where T denotes transpose and

$$D = \begin{bmatrix} \partial I'/\partial I & \partial I'/\partial \Phi \\ \partial \Phi'/\partial I & \partial \Phi'/\partial \Phi \end{bmatrix} , \quad J = \begin{bmatrix} 0 & -1 \\ 1 & 0 \end{bmatrix} . \tag{5}$$

If it exists the generator $G(I, \Phi')$ defines the same map implicitly through the equations

$$I' = I + G_{\Phi'}(I, \Phi') , \quad \Phi = \Phi' + G_I(I, \Phi') , \tag{6}$$

where subscripts denote partial derivatives. By comparison of Eqs. (1) and (2) with Eqs. (6) we see that G must satisfy the partial differential equations

$$G_{\Phi'}(I, \Phi') = R(I, \Phi) , \quad G_I(I, \Phi') = -\Theta(I, \Phi) , \tag{7}$$

or

$$G_{\Phi'}(I, \Phi') = f(I, \Phi') , \quad G_I(I, \Phi') = g(I, \Phi') , \tag{8}$$

where

$$f(I, \Phi') = R(I, \Phi' + F(I, \Phi')) , \quad g(I, \Phi') = F(I, \Phi') . \tag{9}$$

Note that if $G^{(1)}$ and $G^{(2)}$ both satisfy Eqs. (8), then the two solutions differ by a constant at most. Since a constant does not affect the map defined through Eqs. (6), we see that a generator, if any, is essentially unique if F is unique. In Ref.(3) a formula for G was derived as a necessary condition on any solution of Eqs.(8), but there was no proof that the formula actually satisfied all of the equations, and in fact no proof that G exists. One gets more insight, as well as new ideas for computational methods, by first establishing the existence of G.

EXISTENCE OF THE GENERATING FUNCTION

In this section we make minimal assumptions about the given functions R and Θ; namely, that they are in class C^1 (have a continuous first derivative in each of the $2n$ variables), and that Eq.(2) has a unique solution $\Phi = \Phi' + F(I, \Phi')$, where F is in C^1 and is 2π-periodic in Φ'. We also suppose that the Jacobian matrix of the angular map, $1 + \Theta_\Phi$, is nonsingular. These conditions are to hold for I in some open, simply connected set Ω. For the present discussion the angular part of the map given by Eq.(2) is best regarded as a map from $\mathbb{R}^n$ to $\mathbb{R}^n$, although in computations one would usually define angles modulo 2π. In the following sections we shall impose more specific conditions on R, Θ, and Ω, those that arise naturally in an accelerator tracking code. The conditions stated above will then hold automatically.

We seek a solution $G \in C^2$ of Eqs.(7) in the region $\mathcal{D} = \Omega \times \mathbb{R}^n$. Let us define the $2n$-dimensional vector $z = (I, \Phi')$ and write the equations in the form

$$G_z(z) = \gamma(z) . \tag{10}$$

An obvious necessary condition for Eq.(10) to have a solution $G \in C^2$ in an open region is that $\partial \gamma_i / \partial z_j = \partial \gamma_j / \partial z_i$, all i,j; i.e., the tensor equation curl $\gamma = 0$ holds. If that region is also simply connected (as is the region $\mathcal{D}$ in which we work) this condition is sufficient as well. Let ω be the 1-form associated with γ, that is $\omega = \gamma_1 dz_1 + \cdots + \gamma_{2n} dz_{2n}$, and C be a suitably well-behaved curve in $\mathcal{D}$. Then since $\gamma \in C^1$, the curl condition gives $d\omega = 0$ and the generalized Stokes theorem in $2n$ dimensions (Ref.(4),Theorem 6, p.478) gives $\int_C \omega = 0$. It follows that the integral of ω between z_0 and z is independent of path and

$$G(z) = \int_{z_0}^{z} \omega . \tag{11}$$

One sees that (11) satisfies (10) by differentiation, taking account of path independence. As was mentioned above, this solution is unique up to a constant addend.

To complete the proof that G exists, we show that curl $\gamma = 0$ follows from the symplectic condition Eq.(4). In the notation of Eqs. (8), (9) the equations to be verified are

$$f_I = g_{\Phi'}, \quad f_{\Phi'} = f_{\Phi'}^T, \quad g_I = g_I^T, \tag{12}$$

which is to say

$$R_I + R_\Phi F_I = F_{\Phi'}^T , \tag{13}$$

$$R_\Phi (1 + F_{\Phi'}) = (1 + F_{\Phi'}^T) R_\Phi^T , \tag{14}$$

$$F_I = F_I^T . \tag{15}$$

In terms of R and Θ the symplectic conditions are

$$(1 + \Theta_\Phi)(1 + R_I^T) - \Theta_I R_\Phi^T = 1 , \qquad (16)$$
$$R_\Phi(1 + R_I^T) - (1 + R_I)R_\Phi^T = 0 , \qquad (17)$$
$$\Theta_I(1 + \Theta_\Phi^T) - (1 + \Theta_\Phi)\Theta_I^T = 0 . \qquad (18)$$

To get expressions for the derivatives of F, we invoke the definition of F,

$$F(I, \Phi') = -\Theta(I, F(I, \Phi') + \Phi') , \qquad (19)$$

and differentiate and solve to get

$$F_I = -(1 + \Theta_\Phi)^{-1}\Theta_I , \qquad (20)$$
$$F_{\Phi'} = -(1 + \Theta_\Phi)^{-1}\Theta_\Phi . \qquad (21)$$

Now Eq.(15) follows from (20) and (18). To prove Eq.(13), write it with F_I^T replacing F_I, and substitute the derivatives of F from (20) and (21). The result is the same as (16). Finally, to prove Eq.(14), substitute R_I from (13) in (17), and again use the symmetry of F_I. Thus, curl $\gamma = 0$ has been established.

Even without invoking the argument based on Stokes's theorem, one can derive an explicit formula for G, one that obviously satisfies all of the equations (8). Integrate (8) with respect to one variable at a time, using the remaining differential equations and relations (12) to determine the unknown functions of the variables that remain after each integration. For instance, in the case of two degrees of freedom one formula from such a procedure is

$$G(I, \Phi') = \int_0^{\Phi'_1} f_1(I_1, I_2, u, \Phi'_2)du + \int_0^{\Phi'_2} f_2(I_1, I_2, 0, u)du +$$
$$\int_{I_{10}}^{I_1} g_1(u, I_2, 0, 0)du + \int_{I_{20}}^{I_2} g_2(I_{10}, u, 0, 0)du . \qquad (22)$$

This is easily recognized as a path integral of the form (11).

Another formula for G, the one proposed in Ref.(3) and used in all numerical work to date, is based on the Fourier expansion

$$G(I, \Phi') = \sum_m g_m(I) e^{im \cdot \Phi'} . \qquad (23)$$

For all m with at least one non-zero component m_α, the Fourier amplitude may be expressed as

$$g_m(I) = \frac{1}{im_\alpha(2\pi)^n} \int_{T^n} e^{im \cdot \Phi'} R_\alpha(I, \Phi' + F(I, \Phi'))d\Phi'$$
$$= \frac{1}{im_\alpha(2\pi)^n} \int_{T^n} e^{im \cdot (\Phi + \Theta(I, \Phi))} R_\alpha(I, \Phi) \det[1 + \Theta_\Phi(I, \Phi)]d\Phi , \qquad (24)$$

where $T^n = [0, 2\pi]^n$ is the n-torus. To obtain (24) we differentiate (23) with respect to Φ_α, make use of the first equation in (7), and compute $im_\alpha g_m$ by orthogonality in the usual way. In the second expression for the integral, we avoid having to know the function F explicitly by making a change of integration variable $\Phi' \mapsto \Phi$. This is advantageous in numerical computations of the generator, and also convenient in the following analysis. The corresponding expression for $m = 0$ is obtained by integrating differential equations from the other equation of (7), namely

$$\frac{\partial g_0}{\partial I} = -\int_{T^n} \Theta(I, \Phi) \det\left[1 + \Theta_\Phi(I, \Phi)\right] d\Phi . \tag{25}$$

In confirmation of our general arguments one can show by direct computation using Eqs.(12) that Fourier amplitudes of integrals such as (22) agree with g_m as expressed through (24) and (25). The choice of α is correlated with the path, i.e., the order of integrations over single variables. In the course of the calculation one also shows that (22) is 2π-periodic in Φ'.

PROPERTIES OF THE MAP IN POLAR COORDINATES

We now describe more closely the map functions R and Θ that arise from a tracking code built on a symplectic integrator. We first treat the case of betatron motion (oscillations transverse to the beam direction) in one degree of freedom, for which the map in Cartesian coordinates takes the form

$$x' = \cos(2\pi\nu) \, x + \beta \sin(2\pi\nu) \, p + X(x, p) ,$$
$$p' = -\frac{1}{\beta} \sin(2\pi\nu) \, x + \cos(2\pi\nu) \, p + P(x, p) , \tag{26}$$

where ν and β are positive constants (tune, and beta function at the ring position to which the map refers, respectively). Since the full-turn evolution provided by a symplectic integrator amounts to a composition of a large number of polynomial maps, the functions X and P have the form

$$X(x, p) = \sum_{q=2}^{N} X_q(x, p) , \quad P(x, p) = \sum_{q=2}^{N} P_q(x, p) , \tag{27}$$

where X_q and P_q are homogeneous polynomials of degree q. The transformation to canonical polar coordinates (I, Φ) is given by

$$x + i\beta p = (2\beta I)^{1/2} e^{-i\Phi} . \tag{28}$$

Since the map can be written as

$$x' + i\beta p' = e^{-2\pi i\nu}(x + i\beta p) + X + i\beta P , \tag{29}$$

it is easy to derive the following expressions for R and Θ of Eqs.(1) and (2):

$$R(I,\Phi) = [I/2\beta]^{1/2}\left[e^{i(2\pi\nu+\Phi)}(X+i\beta P) + e^{-i(2\pi\nu+\Phi)}(X-i\beta P)\right]$$
$$+ \frac{1}{2\beta}[X^2 + \beta^2 P^2] \,, \tag{30}$$

$$\Theta(I,\Phi) = 2\pi\nu + \frac{1}{2i}\ln\left[\frac{1+e^{-i(2\pi\nu+\Phi)}(X-i\beta P)(2\beta I)^{-1/2}}{1+e^{\,i(2\pi\nu+\Phi)}(X+i\beta P)(2\beta I)^{-1/2}}\right]. \tag{31}$$

At real I and Φ the logarithm is purely imaginary. To be definite we choose the branch to be such that $\Theta - 2\pi\nu \in [-\pi/2, \pi/2]$.

Because X and P are polynomials of second or higher order, the following features are obvious from (28),(30), and (31):

-1- $R(I,\Phi)$ is a polynomial in $I^{1/2}$ and in $\exp(\pm i\Phi)$.

-2- $\Theta(I,\Phi)$ is analytic in $I^{1/2}$ and in $\exp(\pm i\Phi)$ in any region in which $|\exp(\pm i\Phi)(X \pm i\beta P)(2\beta I)^{-1/2}| < 1$.

Now we can discuss inversion of the transformation (2) on $\mathbb{R}$ at real I, an issue in the previous section. Given $\Phi' \in \mathbb{R}$ we show that there is a unique solution $\Phi \in \mathbb{R}$ of (2), provided that I is sufficiently small. We use a method that works as well in the case of two degrees of freedom, even though a simpler argument based on requiring that $\Phi + \Theta(I,\Phi)$ be monotonic can be used in the present case. Let $x = \Phi' - 2\pi\nu - \Phi$, $\Theta(I,\Phi) = 2\pi\nu + T(I,\Phi)$ and write Eq.(2) as a fixed point problem to be solved on $\mathbb{R}$, namely

$$x = A(x)\,, \qquad A(x) = T(I, \Phi' - 2\pi\nu - x)\,. \tag{32}$$

One can see from (31) that $T(I,\Phi)$ and $T_\Phi(I,\Phi)$ are continuous, periodic functions of Φ on $\mathbb{R}$ at sufficiently small I. Since $(X \pm i\beta P)I^{-1/2} = \mathcal{O}(I^{1/2})$, the singularities of the logarithm in (31) are avoided if we choose $\bar{I}$ so that

$$|X(I,\Phi) \pm i\beta P(I,\Phi)|(2\beta I)^{-1/2} \leq \eta < 1\,, \tag{33}$$

for $I < \bar{I}$. (Here we have written $X(I,\Phi)$ for what was previously called $X(x,p)$, and similarly for P.) Moreover, T_Φ has the form $\mathcal{O}(I^{1/2})/(1+\mathcal{O}(I^{1/2}))$ as far as its I-dependence is concerned, and is small for small I, uniformly in Φ. Let us then redefine $\bar{I}$, making it smaller if necessary, so that $|T_\Phi(I,\Phi)| \leq \alpha < 1$, $\Phi \in \mathbb{R}$ for $I < \bar{I}$. It then follows from the contraction mapping theorem that the fixed point problem (32) has a unique solution if $I < \bar{I}$, since $A : \mathbb{R} \to \mathbb{R}$ and $|A(x) - A(y)| \leq \alpha|x-y|$, all $x, y \in \mathbb{R}$, the latter by Taylor's theorem. The corresponding solution Φ of (2) can be written as $\Phi = \Phi' + F(I,\Phi')$, where F is periodic. This follows from (32) and the periodicity of T.

We now have existence and uniqueness of the function F of the previous section, but we also need to know that $F \in C^1$. That may be established by an implicit function argument applied to Eq.(2) written as $H(\Phi, \Phi', I) = 0$. We have already seen that this equation has a unique solution $\Phi(\Phi', I)$ if $I < \bar{I}$. We can conclude, by an appropriate form of the implicit function theorem (Ref.(5), Sections 10.2.2, 10.2.3), that the solution has continuous derivatives in both variables if H has a continuous derivative in each of its three arguments, in a neighborhood of the solution. We have already taken care of $\partial H/\partial \Phi$ by requiring $I < \bar{I}$, and $\partial H/\partial \Phi' = 1$. For $\partial H/\partial I$ we have to add a new requirement on the region, namely that $I > \underline{I} > 0$. This is required since $\partial T/\partial I$ involves a factor $(X \pm i\beta P)I^{-3/2}$ which in general blows up at $I = 0$. In the present case, the region Ω mentioned in the previous section is $\{I | 0 < \underline{I} < I < \bar{I}\}$.

Next consider betatron motion in two degrees of freedom, but with the two motions uncoupled at the linear level. Then the map has the form

$$x'_j = \cos(2\pi\nu_j) x_j + \beta_j \sin(2\pi\nu_j) p_j + X_j(x,p),$$

$$p'_j = -\frac{1}{\beta_j}\sin(2\pi\nu_j) x_j + \cos(2\pi\nu_j) p_j + P_j(x,p),$$

$$j = 1, 2, \tag{34}$$

where now X_j and P_j are sums of homogenous polynomials of degree 2 and greater in x_1, p_1, x_2, p_2. The corresponding map functions, R and Θ, are given by expressions just like (30) and (31), except that all ingredients of the formulas acquire a subscript j, for instance

$$\Theta_j(I, \Phi) = 2\pi\nu_j + \frac{1}{2i}\ln\left[\frac{1 + e^{-i(2\pi\nu_j + \Phi_j)}(X_j - i\beta_j P_j)(2\beta_j I_j)^{-1/2}}{1 + e^{i(2\pi\nu_j + \Phi_j)}(X_j + i\beta_j P_j)(2\beta_j I_j)^{-1/2}}\right]. \tag{35}$$

Now it is clear that claims totally analogous to (–1–) and (–2–) above are valid in the present case; we have polynomial or analytic behavior in $I_j^{1/2}$ and $\exp(\pm i\Phi_j)$, $j = 1, 2$. An important difference arises, however, when we try to verify the condition

$$|\exp(\pm i\Phi_j)(X_j \pm i\beta_j P_j)(2\beta_j I_j)^{-1/2}| \leq \eta < 1. \tag{36}$$

Here the coefficient of $I_j^{-1/2}$ does not necessarily vanish as $I_j \to 0$. It may contain terms like x_k^2, p_k^2, $x_k p_k$ with $k \neq j$, which are proportional to I_k. In general there is a pole in $I_j^{1/2}$ at $I_j^{1/2} = 0$.

Let us see how to deal with this situation when we turn again to the solution of Eq.(2), now in $\mathbb{R}^2$ at real I. We must first find a sufficient condition for inequality (36). For that we ask that I_1 and I_2 be not only small but also not too dissimilar, for instance by requiring

$$I \in \mathcal{K}(\lambda, \mu, \bar{I}) = \{ I \mid \lambda I_1 < I_2 < \mu I_1, \; \|I\| < \bar{I}\},$$
$$0 < \lambda < 1, \quad 1 < \mu < \infty, \tag{37}$$

where $\|\cdot\|$ is the Euclidian norm. Then (36) certainly holds for $\bar{I}$ sufficiently small. Following the plan of the one-dimensional case, we again solve the fixed point problem analogous to (32), taking some vector norm $\|x\|$ and a compatible matrix norm to bound the Jacobian matrix A_x. After a possible downward adjustment of $\bar{I}$ to a new value $\bar{I}_1$, we guarantee that $\|A_x\| \leq \alpha < 1$ for all $I \in \mathcal{K}$, so that a unique solution x is implied by the contraction mapping principle. Again, we have $\Phi = \Phi' + F(I, \Phi')$ with F periodic in Φ'. We can again apply the implicit function theorem to show that $F \in C^1$, provided that we bound $\|I\|$ below to avoid the singularity of $\partial T/\partial I$, which can be more severe by one power than in the one-dimensional case. Thus we require

$$I \in \mathcal{L}(\lambda, \mu, \underline{I}, \bar{I}_1) = \{ I \mid \lambda I_1 < I_2 < \mu I_1 ,\ 0 < \underline{I} < \| I \| < \bar{I}_1 \}, \qquad (38)$$

where $\underline{I}$ can have any positive value less than $\bar{I}_1$. Now $\mathcal{L}$ is the region Ω of the previous section.

CONVERGENCE OF THE FOURIER SERIES

Since R and Θ are analytic in Φ_j, it is natural to use a complex variable method to study the convergence of the series (23). Recall that if a function $f(\phi)$ is 2π-periodic and analytic in a strip $|\text{Im } \phi| < \sigma$, and continuous on the closure of the strip, $|\text{Im } \phi| \leq \sigma$, then its Fourier coefficients obey the bound $|f_m| \leq \|f\| \exp(-|m|\sigma)$. This is seen by distorting the contour in the integral that defines f_m, and applying Cauchy's theorem. For $m > 0$, say, the interval of integration $[0, 2\pi]$ can be replaced by the straight line segment between $-i\sigma$ and $-i\sigma + 2\pi$, since the contributions of vertical paths leading to and from the displaced interval cancel by periodicity. Conversely, if f_m is bounded as stated, then $f(\phi)$ is analytic in the strip, since its Fourier series converges uniformly for $|\text{Im } \phi| \leq \delta < \sigma$. The generalization to a function of several variables, 2π-periodic in each, is obvious.

We give the proof for a map of type (34). As we have already noted, $R(I, \Phi)$ is a polynomial as a function of each $I_j^{1/2}$ and each $\exp(\pm i\Phi_j)$. To discuss analyticity of Θ, we fix σ and suppose that

$$I \in \mathcal{L}(\lambda, \mu, \underline{I}, \bar{I}_2) , \qquad (39)$$

where $\bar{I}_2$ is to be determined, and $\underline{I}$ can have any positive value less than $\bar{I}_2$. Then since X_j and P_j are sums of homogeneous polynomials of degree 2 or greater, there is some $M(\sigma)$ so that

$$\left| e^{\pm i\Phi_j} \left[X_j(I, \Phi) \pm \beta_j P_j(I, \Phi) \right] \right| (2\beta_j I_j)^{-1/2} < M(\sigma) I_j^{1/2} ,$$
$$|\text{Im } \Phi| \leq \sigma . \qquad (40)$$

(We write $|\text{Im }\Phi| \leq \sigma$ to mean $|\text{Im }\Phi_j| \leq \sigma$, $j = 1, 2$). We then choose $\bar{I}_2 \leq \eta M(\bar{\sigma})^{-2}$, where $\eta < 1$. Then $M(\sigma)I_j^{1/2} \leq M(\sigma)\|I\|^{1/2} < \eta < 1$. By (40) and (35) we are then assured that Θ and Θ_Φ are analytic in each Φ_j for $|\text{Im }\Phi| < \bar{\sigma}$ and continuous on $|\text{Im }\Phi| \leq \sigma$.

Now to derive exponential decrease of the Fourier coefficients, we work with Eq. (24), which holds under condition (38). Imposing also condition (39), we can now displace the contour of each integration variable Φ_j in Eq.(24), moving it into the lower (upper) half-plane a distance σ, according as m_j is positive (negative). If any m_j is zero, we need not move the corresponding contour. Notice that we need not be concerned about hitting possible zeros of $\det(1 + \Theta_\Phi)$ at complex Φ. The formula (24) is correct with nonzero determinant at real Φ, and moving the contour to complex Φ is justified by Cauchy's theorem, since the determinant and all other ingredients of the integrand are analytic. To extract exponential decrease of g_m at large m_j, it is sufficient to show existence of a τ such that

$$|\text{Im }\Theta_j(I, \Phi)| \leq \tau < \sigma, \quad |\text{Im }\Phi| \leq \sigma, \quad (41)$$

because then on the complex contour

$$\text{Re}\left[-im_j(\Phi_j + \Theta_j(I, \Phi))\right] \leq -|m_j|(\sigma - \tau). \quad (42)$$

To find a sufficient condition for (41) to hold, we note that

$$\text{Im }\Theta_j = \frac{1}{2}\left[\ln\left|1 + e^{i(2\pi\nu_j + \Phi_j)}(X_j + i\beta_j P_j)(2\beta_j I_j)^{-1/2}\right| - (i \to -i)\right]. \quad (43)$$

Since $\ln(1+x)$ is monotonic and less than x for $x > 0$, we have by Eq.(40) that

$$\left|\text{Im }\Theta_j(I, \Phi)\right| \leq M(\sigma)I_j^{1/2} \leq \tau < \sigma, \quad (44)$$

for $I \in \mathcal{L}(\lambda, \mu, \underline{I}, \bar{I})$ and $|\text{Im }\Phi| \leq \sigma$, for some sufficiently small $\bar{I}$. Let us take $\tau < 1$, so that our previous condition (40) also holds, and denote the resulting value of $\bar{I}$ by $\bar{I}_2$, thus possibly redefining the previous I_2. Now all conditions on I in this and the previous section are met if

$$I \in \mathcal{L}(\lambda, \mu, \underline{I}, \bar{I}), \quad \bar{I} = \min(\bar{I}_1, \bar{I}_2). \quad (45)$$

When (45) holds, we can be sure that g_m decreases exponentially with $|m| = \max_j |m_j|$, and that the same is true for all its I derivatives (modulo powers of $|m|$). The latter is true because we can differentiate the second integral in (24) any number of times, after displacement of the contour, each time bringing down one power of m_j but retaining the exponentially decreasing factor. Thus, under condition (45), all derivatives $g_m^{(i,j)} = \partial g_m^{i+j}/\partial I_i \partial I_j$ are continuous and bounded, and

$$|g_m^{(i,j)}(I)| \leq \kappa_{ij}|m|^{i+j}\exp\left(-|m|(\sigma-\tau)\right), \qquad (46)$$

for some κ_{ij} independent of I, and $i,j = 0, 1, \ldots$. It is not difficult to specify a region in which g_m is analytic as a function of two complex variables I_1, I_2, but we shall omit that discussion in this paper since it is not needed in our applications. We have finished the proof of

Theorem 1: For a system in two degrees of freedom, let the map be as described in Eq.(34). For I in the region $\mathcal{L}(\lambda, \mu, \underline{I}, \bar{I})$, with some sufficiently small $\bar{I}$, the generator $G(I, \Phi')$ exists and is unique up to a constant addend, and its derivatives of any order are continuous and bounded. It is given by a Fourier series (23) that converges absolutely and uniformly for $|\text{Im }\Phi'| \leq \delta$ and $I \in \mathcal{L}(\lambda, \mu, \underline{I}, \bar{I}(\delta))$, for any $\delta > 0$ but with $\bar{I}(\delta)$ tending to zero as δ increases. The same region of convergence occurs for the Fourier series of all derivatives of G, although the convergence may be slower by powers of $|m|$. A suitable $\bar{I}$ can be computed from a knowledge of the map, following the steps of the proof.

A similar statement is of course true for one degree of freedom, in which case $\mathcal{L}$ is replaced by the interval $(\underline{I}, \bar{I})$, $\underline{I} > 0$. Let us illustrate the choice of $\bar{I}$ for the Hénon Map in one degree of freedom, which describes the effect of a thin sextupole magnet together with the rotation in phase space caused by linear forces. The term in the Hamiltonian giving the impulsive sextupole force ("kick") at location $s = 0$ in the ring is $(\lambda/3)x^3\delta(s)$. The kick changes p by $\Delta p = -\lambda x^2$ while leaving x unchanged. The kick followed by the rotation with tune ν gives

$$X - i\beta P = \lambda\beta i e^{2\pi\nu i}x^2,$$

$$\Theta(I,\Phi) = 2\pi\nu + \frac{1}{2i}\ln\left[\frac{1+i\lambda\beta\cos^2\Phi e^{-i\Phi}(2\beta I)^{1/2}}{1-i\lambda\beta\cos^2\Phi e^{i\Phi}(2\beta I)^{1/2}}\right]. \qquad (47)$$

The equation (2) will have a unique solution in $\mathbb{R}$ if $1+\Theta_\Phi$ is positive, and we can guarantee that by requiring

$$\left|\lambda\beta\text{Re}\left[\frac{\partial_\Phi(\cos^2\Phi e^{-i\Phi})(2\beta I)^{1/2}}{1+i\lambda\beta\cos^2\Phi e^{-i\Phi}(2\beta I)^{1/2}}\right]\right| \leq \alpha < 1. \qquad (48)$$

This inequality holds for $I < \bar{I}_1$. Next, to ensure analyticity of Θ and Θ_Φ for $|\text{Im }\Phi| < \sigma$ and continuity in $|\text{Im }\Phi| \leq \sigma$, note first that

$$\left|\lambda\beta(2\beta I)^{1/2}\cos^2\Phi e^{\pm i\Phi}\right| < M(\sigma)I^{1/2},$$
$$|\text{Im }\Phi| \leq \sigma, \quad M(\sigma) = \lambda(\beta/2)^{3/2}(3e^\sigma + e^{3\sigma}). \qquad (49)$$

A nice choice for σ is $\bar{\sigma} = 0.5100\ldots$, the value that maximizes $\sigma/(3\exp\sigma + \exp 3\sigma)$. Choose $\bar{I}_2$ so that $M(\bar{\sigma})(\bar{I}_2)^{1/2} = \tau < \bar{\sigma}$. Then with $\bar{I} = \min(\bar{I}_1, \bar{I}_2)$, a generator G exists, unique up to a constant, and all of its derivatives are continuous and bounded. The Fourier series for G and its derivatives will converge, absolutely and uniformly for $|\text{Im } \Phi| \leq \delta < \bar{\sigma} - \tau$ and $I \in (\underline{I}, \bar{I})$.

Although our sufficient condition for analyticity of Θ is hardly necessary, one can show in the present example that Θ certainly has branch points at sufficiently large Im Φ, regardless of the value of I. With $\Phi = u + iv$ and $\hat{\lambda} = \lambda\beta(\beta I/2)^{1/2}$, the logarithm of (47) has a branch point where

$$\hat{\lambda}[e^{-v}\sin u - e^{3v}\sin 3u - 2e^v \sin u] = 1 ,$$
$$e^{-v}\cos u + e^{3v}\cos 3u + 2e^v \cos u = 0 . \tag{50}$$

With $u = \pi/2$ and $\lambda > 0$ there is a solution where $4\hat{\lambda}e^v \sinh^2 v = 1$; for $\lambda < 0$ take $u = 3\pi/2$.

CONVERGENCE OF THE SPLINE APPROXIMATION

It remains to discuss the approximation of the Fourier coefficients $g_m(I)$ by spline functions of I, of course for I in the region $\mathcal{L}$ specified in the previous section. We need do this only for finite m, say for $|m| < M$, since the various Fourier series converge uniformly. We are mainly interested in the derivatives of G, which occur in the equations (6) that define the map induced by the generator. For fixed $\epsilon > 0$ we choose M so large that

$$\left| G_{\Phi'_j}(I, \Phi') - \sum_{|m|<M} im_j g_m(I) e^{im\cdot\Phi'} \right| < \epsilon/2 ,$$
$$\left| G_{I_j}(I, \Phi') - \sum_{|m|<M} \partial g_m(I)/\partial I_j e^{im\cdot\Phi'} \right| < \epsilon/2 , \tag{51}$$

for $j = 1, 2$ and all $(I, \Phi') \in \mathcal{L} \times T^2$.

Explicit error bounds for approximation of a univariate function and its derivatives by interpolating cubic splines are known for the case of the Hermite boundary conditions, which require that the derivative of the spline match the derivative of the function at the first and last knots (6) . We assume this case, partly to avoid a longer story concerning approximation theorems with more general conditions (7). Numerical computations could be done with Hermite conditions if automatic differentiation (8) (algebra of truncated power series) were used to find first derivatives of the map defined by the tracking code. To date, a different spline definition without derivative data has been used (3, 1) , but it would be interesting to try the Hermite scheme as well.

Given an ascending sequence of spline knots, $x_0 < x_1 < \cdots < x_n$, define $h = \max |x_{i+1} - x_i|$, and $\|f\| = \sup_{x \in [x_0, x_n]} |f(x)|$. Let $s(x)$ be the unique piecewise cubic polynomial function such that $s(x_i) = f(x_i)$, $i = 0, 1, \ldots n$, $s^{(1)}(x_i) = f^{(1)}(x_i)$, $i = 0, n$, and $s \in C^2[x_0, x_n]$. For $f \in C^4[x_0, x_n]$, Hall and Meyer (9) have proved that the following is true, irrespective of the distribution of the knots:

$$\|f^{(i)} - s^{(i)}\| \leq \kappa_i h^{4-i} \|f^{(4)}\|, \quad i = 0, 1, 2,$$
$$(\kappa_0, \kappa_1, \kappa_2) = \left(\frac{5}{384}, \frac{1}{24}, \frac{3}{8}\right). \tag{52}$$

For bivariate spline interpolation of a function $f(x, y)$ we take two knot sequences, $x_0 < x_1 < \cdots < x_n$ and $y_0 < y_1 < \cdots < y_m$, and define $h_x = \max |x_{i+1} - x_i|$, $h_y = \max |y_{i+1} - y_i|$, $\mathcal{R} = [x_0, x_n] \times [y_0, y_m]$, $\|f\| = \sup_\mathcal{R} |f(x, y)|$. Define the operator $\mathcal{P}_x$ so that $\mathcal{P}_x g(x)$ is the cubic spline interpolant of $g(x)$ with Hermite boundary conditions, and similarly for $\mathcal{P}_y$. Then the bicubic spline interpolant of $f(x, y)$ with Hermite boundary conditions is defined to be $s(x, y) = \mathcal{P}_y(\mathcal{P}_x f(x, y)) = \mathcal{P}_x(\mathcal{P}_y f(x, y))$. That is, we first interpolate in one variable, and then interpolate the resulting spline coefficients in the other variable. This is equivalent to expressing $s(x, y)$ in terms of the tensor product basis formed from the cardinal spline basis functions for the two dimensions (6). For $f \in C^4[\mathcal{R}]$, Carlson and Hall (10) showed that, irrespective of knot distribution,

$$\|(f - s)^{(i,j)}\| \leq \epsilon_{4-j,i} h_x^{4-i} \|f^{(4-j,j)}\| + \epsilon_{2i}\epsilon_{2j} h_x^{2-i} h_y^{2-j} \|f^{(2,2)}\|$$
$$+ \epsilon_{4-i,j} h_y^{4-j} \|f^{(i,4-i)}\|, \quad 0 \leq i, j \leq 2, \tag{53}$$

where $g^{(i,j)} = \partial^{i+j} g / \partial x_i \partial x_j$. Values of the $\epsilon_{i,j}$ are given in Table 1 of Ref.(10). In our application we require only the following cases:

$$\|f - s\| \leq \frac{5}{384}\left[h_x^4 \|f^{(4,0)}\| + h_y^4 \|f^{(0,4)}\|\right] + \frac{81}{64} h_x^2 h_y^2 \|f^{(2,2)}\|, \tag{54}$$

$$\|(f - s)^{(1,0)}\| \leq \frac{9 + \sqrt{3}}{216} h_x^3 \|f^{(4,0)}\| + \frac{9}{2} h_x h_y^2 \|f^{(2,2)}\| + \frac{71}{216} h_y^4 \|f^{(1,3)}\|. \tag{55}$$

We wish to approximate $g_m(I_1, I_2)$ by bicubic spline interpolation, on some rectangle $\mathcal{R} = [I_{10}, I_{1n}] \times [I_{20}, I_{2m}] \in \mathcal{L}(\lambda, \mu, \underline{I}, \bar{I})$. For notational convenience, we suppose that the two mesh step bounds are equal and write $h_{I_1} = h_{I_2} = h$. We denote the spline interpolation of a function $f(I)$ by $f^s(I)$. In numerical construction and application of the generator we approximate $g_m(I)$ by $g_m^s(I)$, and then use $\partial g_m^s(I)/\partial I$, calculated analytically, as the approximation to $\partial g_m(I)/\partial I$. If we were to approximate $g_m(I)$ and $\partial g_m(I)/\partial I$ independently by cubic splines, then the symplectic condition would not be maintained exactly, and the whole rationale of the generating function method would be undermined. Thus, we shall find an application for (55) as well as (54).

By applying Eqs. (46, 51, 54, 55), we bound the errors for the final approximations to $G_{\Phi'}$ and G_I as follows:

$$\left|G_{\Phi'_j}(I,\Phi') - \sum_{|m|<M} im_j g^s_m(I)e^{im\cdot\Phi'}\right| \leq$$

$$\left|G_{\Phi'_j}(I,\Phi') - \sum_{|m|<M} im_j g_m(I)e^{im\cdot\Phi'}\right| + \left|\sum_{|m|<M} im_j\big((g_m(I) - g^s_m(I))\big)e^{im\cdot\Phi'}\right| \leq$$

$$\frac{\epsilon}{2} + h^4\left(\frac{5}{384}(\kappa_{40} + \kappa_{04}) + \frac{81}{64}\kappa_{22}\right)\sum_{|m|<M} |m|^5 e^{-|m|(\sigma-\tau)} , \qquad (56)$$

$$\left|G_{I_1}(I,\Phi') - \sum_{|m|<M} g^{s\,(1,0)}_m(I)e^{im\cdot\Phi'}\right| \leq$$

$$\left|G_{I_1}(I,\Phi') - \sum_{|m|<M} g^{(1,0)}_m(I)e^{im\cdot\Phi'}\right| + \left|\sum_{|m|<M} \big((g^{(1,0)}_m(I) - g^{s\,(1,0)}_m(I))\big)e^{im\cdot\Phi'}\right| \leq$$

$$\frac{\epsilon}{2} + h^3\left(\frac{9+\sqrt{3}}{216}\kappa_{40} + \frac{9}{2}\kappa_{22} + \frac{71}{216}h\kappa_{13}\right)\sum_{|m|<M} |m|^4 e^{-|m|(\sigma-\tau)} , \qquad (57)$$

and similarly for G_{I_2}. Each of the right hand sides can be made less than ϵ, by taking $h = \mathcal{O}(\epsilon^{1/3})$ sufficiently small. This completes the proof of

Theorem 2: Let G be the generating function for the map (34) with $I \in \mathcal{L}$, as described in Theorem 1. The Fourier series for G can be approximated by the series $G^{(h,M)}$, which is obtained by truncating the series for G at $|m| = M$, then approximating the coefficients by bicubic spline interpolation with Hermite boundary conditions on a rectangle $\mathcal{R} \in \mathcal{L}$, the parameter h being the maximum mesh step in either direction. For sufficiently large $M = M(\epsilon)$ and sufficiently small $h = \mathcal{O}(\epsilon^{1/3})$, the series $G^{(h,M)}_{\Phi'}$ and $G^{(h,M)}_I$ approximate $G_{\Phi'}$ and G_I within an error ϵ, uniformly for $(I,\Phi') \in \mathcal{R} \times T^2$.

COMMENTS

We have seen that elementary arguments prove convergence of the Fourier-spline representation of the generating function of a full turn map as defined by a symplectic tracking code. As is usual in analyses of this sort, the specific estimates for the region of validity of the Fourier-spline series are probably somewhat pessimistic from a practical stand point. Nevertheless, the analysis reveals the analytic structure of the generating function and gives rates of convergence, results that should be useful in a search for improvements in the practical realization of the method. One feature of the proof shows up very clearly in numerical work (1), (3), namely the restriction to a region in the I_1, I_2 plane that excludes neighborhoods of the coordinate axes. A high priority for further work is to avoid this problem, which is really a question of a coordinate singularity, by using Cartesian coordinates. One possibility is a straightforward adaptation of the present method, replacing the Fourier development by an expansion in Hermite polynomials.

ACKNOWLEDGMENTS

We wish to thank Dr. Zohreh Parsa, Coordinator of the Conference on Particle Beam Stability and Nonlinear Dynamics, and Prof. James Hartle, Director of the Institute of Theoretical Physics, Santa Barbara, for the opportunity to attend a most interesting and enjoyable meeting. The work of R. L. Warnock was supported in part by U. S. Department of Energy contract DE–AC03–76SF00515.

REFERENCES

1. Warnock, R. L., and Berg, J. S., in *Proceedings of the ICFA Workshop on Nonlinear and Collective Phenomena in Beam Physics, Arcidosso, Italy, September 2-6, 1996*, to be published in AIP Conference Proceedings.
2. Warnock, R. L., and Berg, J. S., SLAC-PUB-95-7045, 1995, to be published in *Proceedings of NATO Advanced Study Institute on Hamiltonian Systems with Three or More Degrees of Freedom*, Catalunya, Spain, June 19-30, 1995.
3. Berg, J. S., Warnock, R. L., Ruth, R. D., and Forest, É., *Phys. Rev. E* **49**, 722 (1994).
4. Buck, R. C., *Advanced Calculus*, New York: McGraw-Hill, Third Edition, 1978.
5. Dieudonné, J., *Foundations of Modern Analysis*, New York: Academic Press 1960.
6. Ahlberg, J. H., Nilson, E. N., and Walsh, J. L., *The Theory of Splines and Their Applications*, New York: Academic Press 1967.
7. de Boor, C., *A Practical Guide to Splines*, New York: Springer, 1978.
8. Rall, L. B., *Automatic Differentiation: Techniques and Applications, Lecture Notes in Computer Science* **120**, Berlin: Springer, 1981.
9. Hall, C. A., *J. Approx. Theory* **1**, 209 (1968); Hall, C. A. and Meyer, W. W., *ibid.* **16**, 105 (1976).
10. Carlson, R. E., and Hall, C. A., *J. Approx. Theory* **7**, 41 (1973).

Algorithms for the Treatment and Analysis of Spin Dynamics in SU(2) and SO(3)

Vladimir Balandin[*,+], Martin Berz[*] and Nina Golubeva[+]

[*]*Department of Physics and Astronomy and National Superconducting Cyclotron Laboratory Michigan State University, East Lansing, MI 48824*
[+]*Institute for Nuclear Research of RAS 60th October Anniversary Pr., 7a Moscow, Russia*

Abstract

A method is described that allows the computation and analysis of high-order spin maps for general non-autonomous optical systems. It is shown how the equations of motion in curvilinear coordinates resulting from the Thomas-BMT equation can be solved within a differential algebraic framework in SU(2) and SO(3) representations.

The resulting maps are subjected to a spin-orbit normal form transformation, and the nonlinear orbit dependencies of the invariant polarization axis as well as the orbit dependent spin tune can be obtained. For the case of electron machines, the resulting invariant polarization can be used to determine the radiative equilibrium polarization via the Derbenev-Kondratenko approach. Both the computation of the spin map as well as the algorithm for the computation of the invariant axis have been implemented in the code COSY INFINITY [2] [3] [4].

THE ONE TURN MAP FOR SPIN-ORBIT MOTION

When viewing the motion "stroboscopically" at a fixed point in the ring (for example, at the point of physics experiment), it is convenient to view and analyze the motion in terms of a one-turn map instead of in terms of the differential equations of motion.

The equations of spin-orbit motion are linear in the spin, and hence the transformation of the spin variables can be described in terms of a matrix, the elements of which depend on the orbital quantities only. The orbital quantities themselves are unaffected by the spin motion, such that altogether the map has the form

$$\vec{x}_f = \mathcal{M}(\vec{x}_i)$$
$$\vec{s}_f = A(\vec{x}_i) \cdot \vec{s}_i$$

where $A(\vec{x}) \in SO(3)$.

The practical computation of the spin-orbit map can be achieved in a variety of ways. Conceptually the simplest way is to interpret it as a motion in the nine variables consisting of orbit and spin. In this case, the DA method allows the computation of the spin-orbit map in the two conventional ways, namely via a propagation operator for the case of the z-independent fields like main fields, and via integration of the equations of motion with DA [5] [7]. But in this simplest method, the number of independent variables increases from six to nine, which particularly in higher orders entails a rather substantial increase of computational and storage requirements. This limits the ability to perform analysis and computation of spin motion to high orders.

Due to the special structure of the equations of motion, it is possible to rephrase the dynamics such that it is still described in terms of only the six orbital variables. For this purpose, we derive the equation of motion for the individual elements of the matrix $A(\vec{x})$. To this end, differential equations for the coefficients of the matrix A are set up, which are integrated along with the differential equations for orbit motion. In practice, the orthogonal symmetry also may be used to save computational expense.

INVARIANT FUNCTIONS AND THE STABLE DIRECTION OF POLARIZATION

We call a function $V(\vec{x}, \vec{s})$ an invariant function of the motion if

$$V(\vec{x}, \vec{s}) = V(\mathcal{M}(\vec{x}), A(\vec{x}) \cdot \vec{s})$$

Considering that in the motion the orbital part is independent of the spin motion as well as the fact that the motion is linear in the spin variables, it is sufficient to consider only functions of the special form

$$V(\vec{x}, \vec{s}) = b(\vec{x}) + \vec{g}(\vec{x}) \cdot \vec{s}.$$

In fact, substituting the expression in the condition of invariance, it follows that b is the usual invariant function of the orbital part of the map, and the search can be restricted to

$$V(\vec{x}, \vec{s}) = \vec{g}(\vec{x}) \cdot \vec{s}.$$

Inserting this into the transfer map, we obtain the necessary condition

$$A(\vec{x}) \cdot \vec{g}(\vec{x}) = \vec{g}(\mathcal{M}(\vec{x})).$$

In the case $\vec{g}(\vec{0}) \neq 0$, we may even scale the function in such a way that $|\vec{g}(\vec{x})| = 1|$. If the matrix A is not the identity, then this g is even unique up to multiplication with invariant functions of the orbital map. The quantity $\vec{g}$ has a rather straightforward interpretation, as it means that the projection of the spin vector on $\vec{g}$ is conserved, and thus it defines an axis along the polarization is conserved.

It has been recognized duly that the existence of an invariant function is highly important, as in practice it allows the injection of a polarized beam along $\vec{g}$ and its preservation for long times.

An advantage of this definition of $\vec{g}$, which was introduced in [6], does not depend on the Hamiltonian form or the coordinate system or any other specifics of the orbit motion as in the original paper of Derbenev and Kondratenko [1].

NORMAL FORM ALGORITHM FOR THE SPIN-ORBIT MAP

Introduce new variables $\vec{y}$ and $\vec{\xi}$ via

$$\vec{y} = \mathcal{K}(\vec{x}), \vec{\xi} = C(\vec{x}) \cdot \vec{s},$$

where $C(\vec{x}) \in SO(3)$. In these variables, the map has the form

$$\vec{y}_f = \mathcal{K}(\mathcal{M}(\mathcal{K}^{-1}(\vec{y}_i))) = \mathcal{N}(\vec{y}_i)$$
$$\vec{\xi}_f = C(\mathcal{M}(\mathcal{K}^{-1}(\vec{y}_i))) \cdot A(\mathcal{K}^{-1}(\vec{y}_i)) \cdot C^{-1}(\mathcal{K}^{-1}(\vec{y}_i)) \cdot \vec{\xi}_i = A(\vec{y}_i) \cdot \vec{\xi}_i$$

The goal is now to show that if there are no resonances up to order m, then the coordinate transformation may be chosen in such a way that the matrix A will up to order n have the form of a simple rotation

$$A(\vec{x}) = \begin{pmatrix} \cos(\lambda(\vec{x})) & \sin(\lambda(\vec{x})) & 0 \\ -\sin(\lambda(\vec{x})) & \cos(\lambda(\vec{x})) & 0 \\ 0 & 0 & 1 \end{pmatrix},$$

and in the new variables, the invariant function is simply $V = \xi_3$. The procedure to obtain this form consists of two steps, first a normal form transformation for the orbit part, and then a subsequent transformation for the spin part.

DA Normal Form Algorithm for the Orbital Map

The first process of successive coordinate substitutions to obtain the invariant functions requires to perform the transformation of the orbital map. The starting step consists of the fixed-point transformation and the linear diagonalization. All further steps are purely nonlinear and no longer affect the linear part.

We begin the mth step by splitting the momentary map $\mathcal{M}$ into its linear and nonlinear parts $\mathcal{R}$ and $\mathcal{S}_m$. Then we perform a transformation using a map $\mathcal{K}_m = \mathcal{I} + \mathcal{T}_m$, where $\mathcal{T}_m$ vanishes to order $m - 1$. To study the effect of the transformation, we now infer up to order m:

$$\mathcal{K} \circ \mathcal{M} \circ \mathcal{K}^{-1} =_m \mathcal{R} + \mathcal{S}_m + (\mathcal{T}_m \circ \mathcal{R} - \mathcal{R} \circ \mathcal{T}_m).$$

A close inspection of the equation reveals that S_m can be simplified by choosing the commutator $\{T_m, \mathcal{R}\} = T_m \circ \mathcal{R} - \mathcal{R} \circ T_m$ appropriately. The detailed description of the procedure can be found in [7,8].

The SU(2) Representation of the Spin Map

To reduce the calculations we note that while the orthogonal 3 x 3 matrix $A(\vec{x}) \in SO(3)$, which defines the spin part of the map, consists of 9 elements, it can be described completely using a smaller number of parameters. We will realize it using the connection between the $SO(3)$ and $SU(2)$ groups. Any matrix $U \in SU(2)$ has the form

$$U = \begin{pmatrix} a & b \\ -b^* & a^* \end{pmatrix}, \text{ where } a \cdot a^* + b \cdot b^* = 1.$$

Corresponding to the vector $\vec{s}$, we define a matrix

$$L = \begin{pmatrix} s_3 & s_1 + is_3 \\ s_1 - is_2 & -s_3 \end{pmatrix}$$

and represent the map in the form

$$\vec{x}_f = \mathcal{M}(\vec{x}_i)$$
$$L_f = U(\vec{x}_i) \cdot L_i \cdot U^*(\vec{x}_i)$$

where $U(\vec{x}) \in SU(2)$ is given as above. Simple arithmetic reveals the following connection between the matrices A and U via

$$A = \begin{pmatrix} \operatorname{Re}(a^2 - b^2) & -\operatorname{Im}(a^2 + b^2) & -2\operatorname{Re}(ab) \\ \operatorname{Im}(a^2 - b^2) & \operatorname{Re}(a^2 + b^2) & -2\operatorname{Im}(ab) \\ 2\operatorname{Re}(ab^*) & 2\operatorname{Im}(ab^*) & aa^* - bb^* \end{pmatrix}$$

Normal Form Algorithm for the SU(2) Spin Map

We begin from the spin-orbit map in the form

$$\vec{x}_f = \mathcal{M}(\vec{x}_i)$$
$$L_f = U(\vec{x}_i) \cdot L_i \cdot U^*(\vec{x}_i),$$

where $\mathcal{N}(\vec{x})$ is the normal form of the orbital part of the map up to order m. Consider the coordinate transformation from the old matrix L to the new matrix $\bar{L}$ by the equation

$$L = C(\vec{x}) \cdot \bar{L} \cdot C^*(\vec{x}), \text{ where } C(\vec{x}) \in SU(2).$$

In the new variables, the map has the form

$$\vec{x}_f = \mathcal{N}(\vec{x}_i)$$
$$\bar{L}_f = \bar{U}(\vec{x}_i) \cdot \bar{L}_i \cdot \bar{U}^*(\vec{x}_i),$$

where $\bar{U}(\vec{x}) = C^*(\mathcal{N}(\vec{x})) \cdot U(\vec{x}) \cdot C(\vec{x})$. If there are no resonances between orbital and spin tunes up to order m, we can find a matrix $C(\vec{x})$ such that the matrix $\bar{U}(\vec{x})$ will be diagonal up to order m and will have the form

$$\bar{U}(\vec{x}) =_m \text{diag}(\exp(i\kappa(I)), \exp(-i\kappa(I))),$$

where I are the invariants of orbital motion. In this case,

$$A(\vec{x}) =_m \begin{pmatrix} \cos(\lambda(\vec{x})) & \sin(\lambda(\vec{x})) & 0 \\ -\sin(\lambda(\vec{x})) & \cos(\lambda(\vec{x})) & 0 \\ 0 & 0 & 1 \end{pmatrix},$$

and the approximate invariant function and spin tune are

$$V = s_3, \quad \lambda(I) = 2\kappa(I).$$

Acknowledgments

For financial support, we are grateful to the US Department of Energy, Grant No. DE-FG02-95ER40931, and the Alfred P. Sloan Foundation

References

1. Derbenev, Ya. S.., and Kondratenko, A. M.., Sov. Phys. JETP. 35 (1972) 230, 37 (1973) 968.

2. Berz, M., COSY INFINITY Version 6 reference manual. Technical Report MSUCL-869, National Superconducting Cyclotron Laboratory, Michigan State University, East Lansing, MI 48824,1993.

3. Berz, M., COSY INFINITY Version 6. In Berz, M., Martin, S., and Ziegler, K., (Eds.), Proc. Nonlinear Effects in Accelerators, page 125. IOP Publishing, 1992.

4. Berz, M., New features in COSY INFINITY. In Third Computational Accelerator Physics Conference, page 267. AIP Conference Proceedings 297,1993.

5. Berz, M., Differential algebraic description of beam dynamics to very high orders. Particle Accelerators, 24:109,1989.

6. Balandin, V., and Golubeva, N., in Proc. of the 1993 Part. Accel. Conf., Washington, 1993.

7. Berz, M., High-Order Computation and Normal Form Analysis of Repetitive Systems, in: M. Month (Ed), Physics of Particle Accelerators, volume AIP 249, page 456. American Institute of Physics, 1991.

8. Berz, M., Differential algebraic formulation of normal form theory. In M. Berz, S. Martin and K. Ziegler (Eds.), Proc. Nonlinear Effects in Accelerators, page 77. IOP Publishing, 1992.

NORMAL FORM ANALYSIS OF THE LHC DYNAMIC APERTURE *

F. Schmidt, CERN, Geneva, Switzerland and
E. Todesco, INFN, Sezione di Bologna

Abstract

Normal form is a well developed tool to study the nonlinear dynamics of even the most complicated structures. This can become particularly useful in the design of large nonlinear hadron colliders like the LHC. In this study the correlation of various quality factors derived through normal forms with the dynamic aperture are used to investigate a machine that is dominated by octupolar errors.

1 Introduction

Almost ten years ago the normal form technique [1-5] has been introduced to the accelerator community. Although this technique allows an in-depth analysis of complicated structures like the Large Hadron Collider (LHC), its acceptance as a standard tool has been rather slow. This is mainly due to the rather abstract character of the theory; indeed, we believe that this is also due to the lack of easy-to-use tools with a simple interface to existing codes. Normal forms codes written by independent groups (i.e., the DaLie code by Forest [6] and the ARES code by Bazzani et al. [7,8]) have been used at CERN to analyse the nonlinear dynamics in the LHC [9,10]. Recently, the NERO program [11] has been developed on the basis of well-tested existing tools [8,12], in order to allow an easy analysis of resonances (including contributions from higher orders) and a simple interface with tracking codes. The user-friendliness is ensured by allowing the analysis of internally or externally produced maps and by providing an elaborate user's manual [13]. For the time being the program reads maps as produced by SIXTRACK [14] which uses the differential algebra package of Berz [3] (on request the program can be adapted to any other map format).

The main advantage of the analysis based on maps and perturbative tools that can be carried out through NERO or similar codes is the possibility of analysing the correlations of the dynamic aperture with some quantities, called quality factors, that can be computed in a relatively easy way. It is well-known that in each lattice the dynamic aperture is dominated by some effects, and therefore some quality factors can be well-correlated and some not. Indeed, the correlation of a given quality factor indicates what is the dominant effect, and therefore suggests a way to correct it.

Both NERO and the DaLie codes have been used in this report to study 60 different realizations of the errors of the latest LHC lattice (version 4). In the first section the quality of the correlation of the dynamic aperture with two simple quality factors are tested. In the second section more sophisticated tools based on the computation of the detuning and on the resonance strength are used. Results about the correlation of resonance norms including higher order contributions are presented in the third section.

This research was supported in part by the National Science Foundation under Grant No. PHY94-07194.

It has to be mentioned that for all studies with NERO the maximum amplitude where the quality factors are evaluated is fixed to a value of eight beam sigmas which is close to the lower end of the dynamic aperture. It must be pointed out that one has to select an amplitude in order to weight the contributions of the higher orders with respect to the lowest one: if one restricts the analysis to one order, no amplitude consideration is necessary.

2 Simple quality factors

The dynamic aperture of LHC version 4 is dominated by just two values [15], i.e. the octupole bias b_4 and a_4 of the main dipoles which are caused by fabrication tolerances. In [15] a correlation of the dynamic aperture has been found with a octupole strength norm

$$\sqrt{\frac{b_4{}^2 + a_4{}^2}{2}}. \qquad (1)$$

In Tab.1 the correlation coefficient R^2 of this case is denoted by 'I'. Although the correlation is not bad we have hoped to get an improvement by a more refined study. Moreover, one may need something else as the knowledge of the multipolar errors may not be available with sufficient precision to power the correction circuits.

Table 1: *Correlation coefficient R^2 for all studied cases*

#	Case	Description	Correlation Coefficient R^2			
			Map Order			
			2	3	5	7
I	Octupole Strength Norm		0.61			
II			$3 \cdot 10^{-4}$			
IIa	Map			0.63		
IIb	Norm					0.64
IIc		after Correction				0.12
IIIa	Tune			0.25		
IIIb	Norm				0.42	
IIIc						0.41
IVa	First	(2 –2)		0.25		
IVb	Order	(3 1)		0.12		
IVs	Resonance	(1 –1)		0.22		
IVd	Strength	Combination		0.64		
Va	High	(2 –2)		0.25	0.25	0.39
Vb	Order	(3 1)		0.12	0.23	0.46
Vc	Resonances	Combination		0.33	0.42	0.52

A very simple quality factor can be defined as the norm of the nonlinear part of the map, that gives a rough indication about the strength of the nonlinearities in a given mapping. It is defined as the sum of the absolute values of all map coefficients $|F_{i,n}|$ relative to the order n, weighted with the power of the amplitude A (see Ref. [10]):

$$\sum_n A^n \cdot \left(\sum_i |F_{i,n}| \right). \tag{2}$$

Fig. 1 and 2 show the correlation of the map norm with the dynamic aperture. While at order 2 there is no correlation, at order three, where the first-order contribution of the octupoles is present, the correlation is good, and slightly increases up to order seven. Both map norms up to order three and up to order seven feature a correlation that is slightly better than the octupole strength norm (0.63 and 0.64 respectively, see Table 1). In fact, we also evaluated the correlation of the map norm up to order three with the octupole norm, finding a good correlation as expected (see Fig. 3). This implies that the sextupoles which also feed the map coefficients of order three can be neglected.

After the large octupolar bias values are corrected, the map norm is practically not correlated any more ($\mathbf{R}^2 = 0.12$) with the improved dynamic aperture. This means that after the correction the dynamic aperture is not dominated by the large octupolar terms as before, but is determined by more intricate effects that become relevant when the octupole bias is wiped out.

As the map norm is excellently correlated with the octupole strength norm but also well correlated with the dynamic aperture it can be used to correct the octupole effect. To this end a map has been produced with the strength of b_4 and a_4 spool piece correctors placed in the magnet ends as a parameter. The third order coefficients and thereby the third order map norm are therefore known as a function of the two corrector families. Using the first order terms only, the corrector strengths required to zero the map norm can easily be predicted. In Fig. 4 the original bias of the b_4 coefficients of the 60 error representations is compared with the resulting b_4 bias after the map norm cancellation with spool pieces. The improvement of dynamic aperture after such a correction has not been tested but it will probably be very beneficial given the good correlations between the map norm and the dynamic aperture. The possibility to predict a correction for a dominant multipole coefficient after installation of the machine is very attractive, in particular as not all magnets may have been measured. The prerequisites are a sufficiently large number of correctors installed in the machine and some means to measure the map. One possible candidate to measure those map coefficients is the recently proposed method to derive them from the spectrum lines of a kicked pencil beam [16].

3 Tune Norm and First Order Resonances

The previous quality factors are based on the simple inspection of the octupole strength and map coefficients respectively. Normal forms provide a perturbative way to compute both detuning and resonances. Nonresonant normal form provide the tune as a function of amplitude, and one can compute the average of the tune at a fixed amplitude A (that has the same meaning as before) in a analytical way, by using the perturbative series. We refer to [10] for more details.

It is now rather well understood that the tuneshift, as any other quality factor, is not *a priori* correlated with the dynamic aperture. Indeed, it strongly depends on the particular lattice if the dynamic aperture is dominated by the detuning, or by some resonance, or simply by the norm of the map, as it seems to be case here. The correlation of the detuning with the dynamic aperture, even though it is increasing with the order, is always worse than the correlation with the norm of the map (case III in Tab. 1). This provides some relevant information: the detuning is not the driving mechanism of the dynamic aperture.

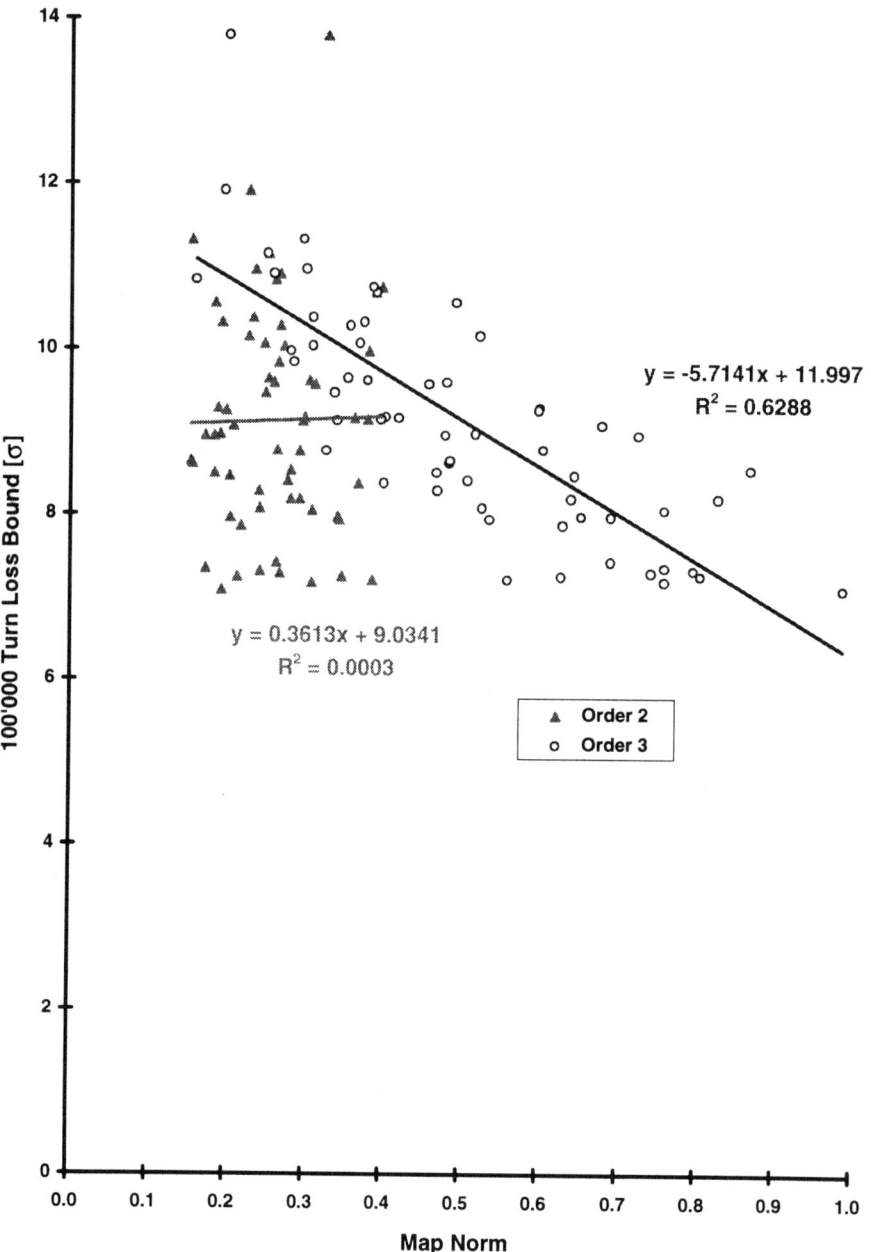

Figure 1: *Correlation of the dynamic aperture with the map norm of order two and three*

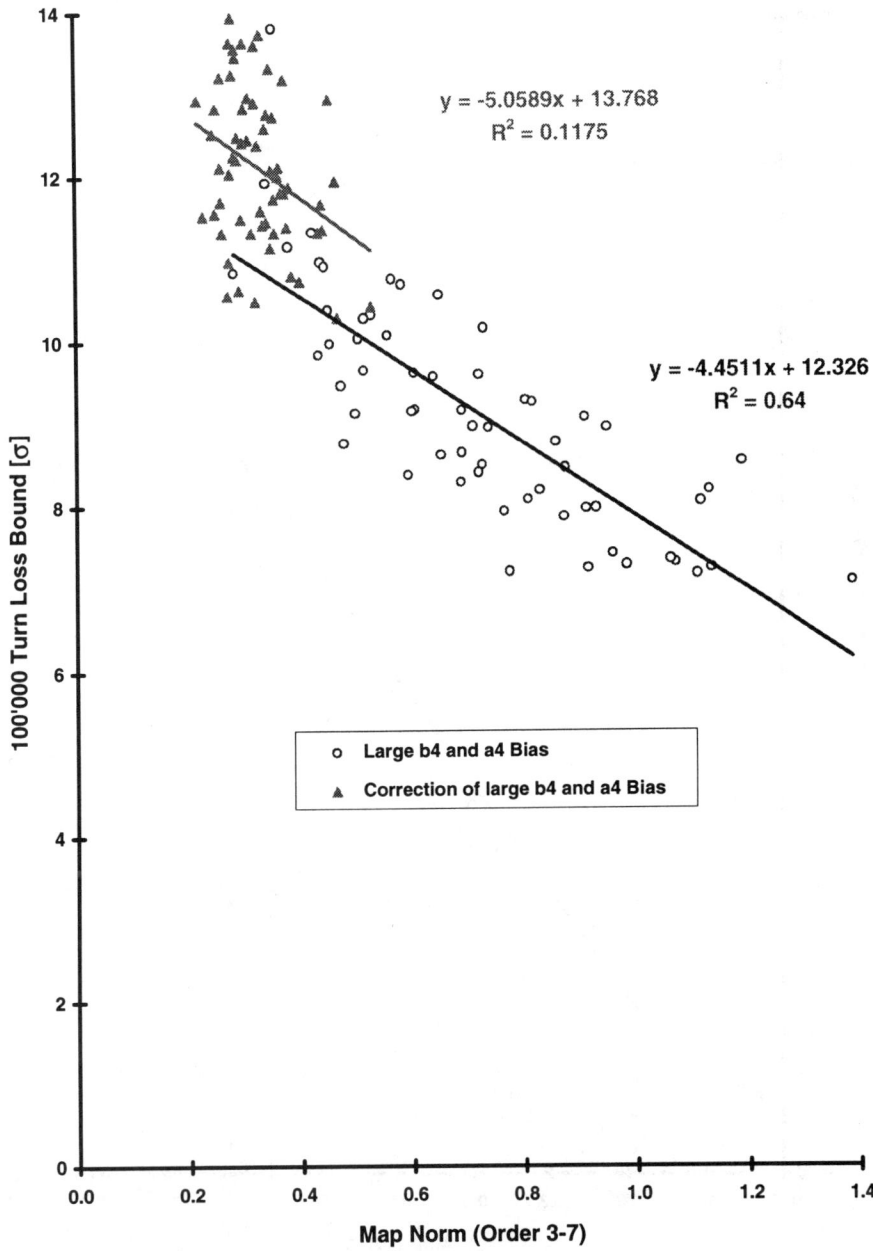

Figure 2: *Correlation of the dynamic aperture with the map norm (order three through seven) before and after the correction of the large b_4 and a_4 bias*

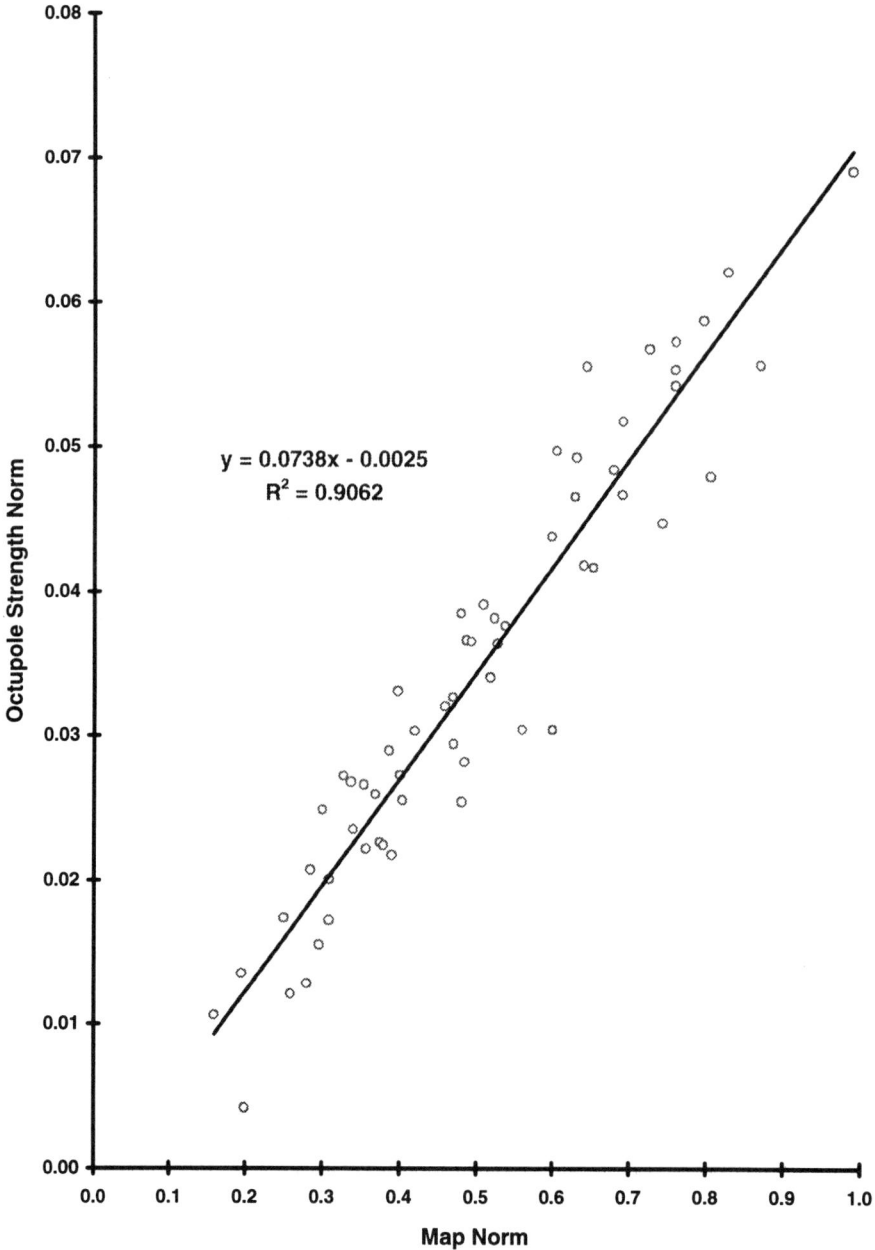

Figure 3: *Correlation of the octupole strength norm with the map norm (order three)*

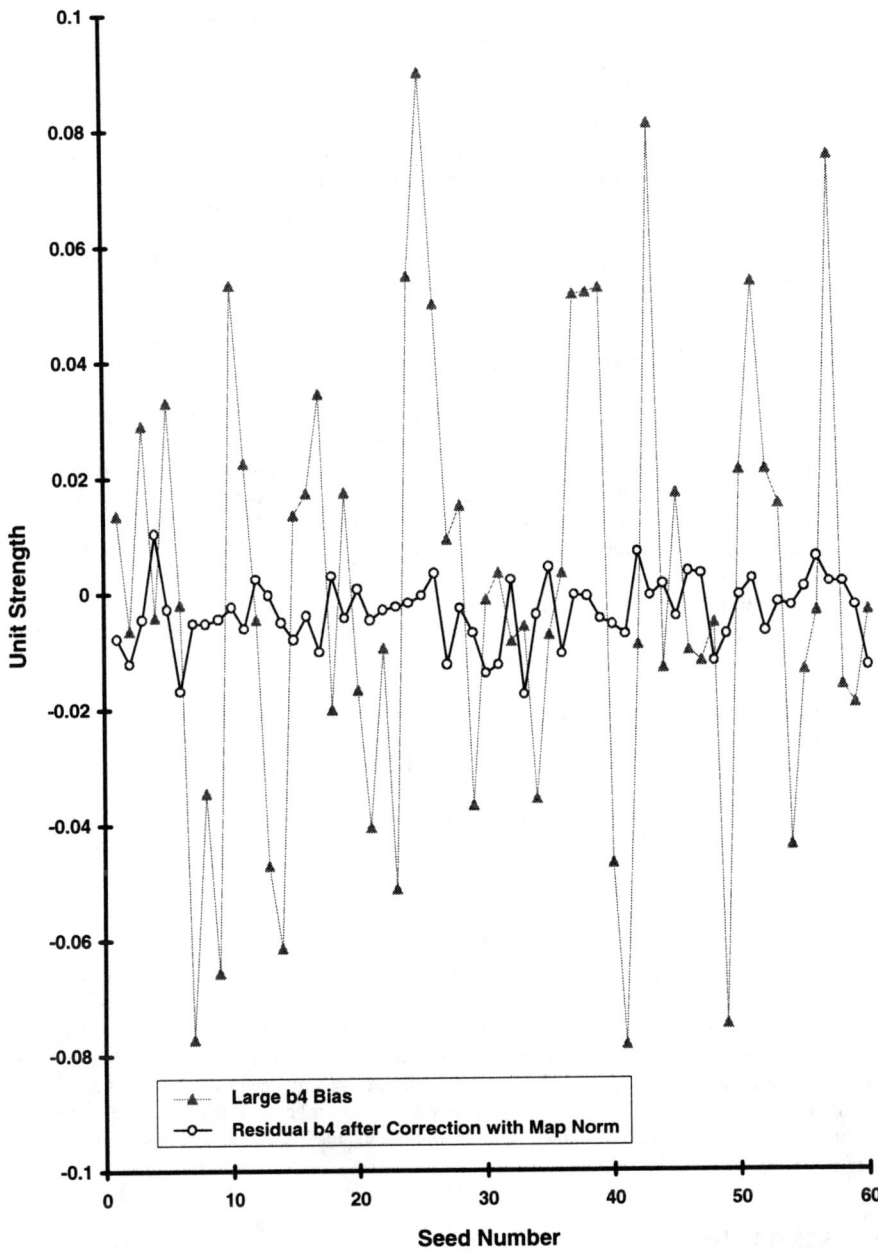

Figure 4: *Correction of the b4 bias using the map norm*

Even more interesting is the study of resonances. For the same machine realizations considered here first order octupole resonances [1] have been found to be correlated with the dynamic aperture [17]. The authors have discovered that the linear multi-variant fit of the three largest resonances, i.e. the (2 –2), the (3 1) resonance and the first order sub-resonance (1 –1) [18] lead to a correlation factor of $R = 0.8$. With the DaLie program we have calculated the same cases except that we use the Hamiltonian H instead of the conjugating function F (both in resonance basis). Although the results should be equivalent to good approximation we prefer the Hamiltonian as it is the starting point for the higher order study to follow below. Moreover, the Hamiltonian reproduces the same phases that one expects from first order calculations à la Guignard [18]. For the linear multi-variant fit we use the $LINEST$ program of $EXCEL$ and of course the results agree with those of Ref. [17] (see part 'IV' in Tab. 1).

4 Higher Order Resonances

The NERO program mainly aims at calculating resonance at higher orders. The resonant perturbative formalism is carried out at arbitrary order for generic mappings. A norm of the resonant part of the interpolating Hamiltonian is worked out. When only the first order contribution is considered, this corresponds to the standard first order perturbative theory based on Hamiltonians [18]. One of the main advantages of the code is that it allows to determine whether the higher orders are relevant or not, thus providing relevant information on the different effects that contribute to the determination of the dynamic aperture.

For example, the correlation between the norm of the (3 1) resonance and the dynamic aperture is shown in Fig. 5 up to order three, five and seven. The correlation coefficient R^2 jumps by some factor of four comparing order seven with three: the higher orders seem to be rather relevant in this case. The figure shows that this improvement is due to a de–population of the lower left corner of the figure. This procedure has also been performed for the (2 –2) resonance. Also in this case we have a correlation improvement, even though it is less pronounced. Nevertheless, with two resonances only we have achieved a correlation coefficient better than 0.5 at order seven (see part 'V' in Tab. 1). Of course one would have to check if the addition of the (1 –1) resonance would further improve the correlation quality. Unfortunately resonances of order two such as the (1 –1) are not yet considered in the code which analyses presently only resonances of order greater or equal to three.

5 Conclusion

Various norms were found to be correlated with the dynamic aperture of 60 error representations of the LHC lattice version 4. The octupole strength norm, the map norm and the linear multi-variant fit of three first order resonances all have a correlation coefficient R^2 slightly above 0.6. There are also indications that the higher order analysis of resonances may improve the correlation quality. As the map norm is the simplest and may be obtained in the easiest way it appears to be the best choice for this lattice lay-out. Moreover, it has been shown that it can be used to predict the correction strength for the bias of large b_4 and a_4. The usefulness of this correction will have to be shown experimentally.

6 Acknowledgements

We would to thank M. Böge for helping us with software tools to evaluate the maps.

---

[1] Of course these resonance are also driven by sextupoles in second order. For our purpose here there is no need to separate the effects because the sextupole contributions are small following the argument of above.

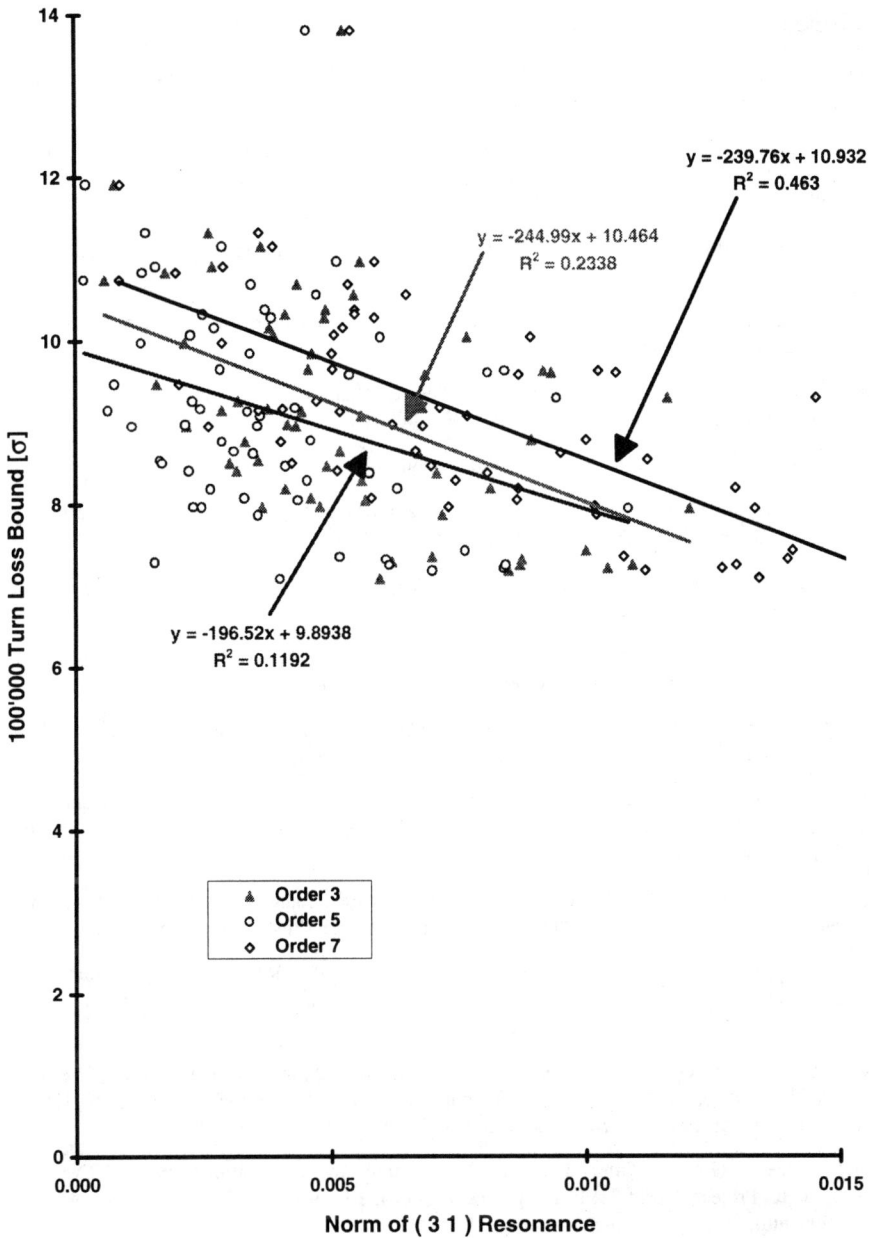

Figure 5: *Correlation of the dynamic aperture with the resonance norm for the (3 1) resonance between order three and seven*

References

[1] A. Bazzani et al., "Normal forms for Hamiltonian maps and nonlinear effects in a particle accelerator", Nuovo Cim., B **102**, pp. 51–80 (1988).

[2] M. Berz, É. Forest and J. Irwin, "Normal form methods for complicated periodic systems: a complete solution using differential algebra and lie operators", Part. Acc. **24**, pp. 91–107 (1989).

[3] M. Berz, "Differential-algebraic description of beam dynamics to very high orders", Part. Acc. **24**, pp. 109–124 (1989).

[4] A. Bazzani et al. , "Resonant normal forms, interpolating Hamiltonians and stability analysis of area preserving maps", Physica D **64**, pp. 66–93 (1993).

[5] A. Bazzani, E. Todesco, G. Turchetti and G. Servizi, "A normal form approach to the theory of nonlinear betatronic motion", CERN 94–02.

[6] É. Forest, written in 1986, private communication.

[7] G. Servizi and G. Turchetti, "A computer program for the Birkhoff series for area preserving maps", Comput. Phys. Commun., **32**, pp. 201–7 (1984).

[8] A. Bazzani, M. Giovannozzi, E. Todesco, "A program to compute Birkhoff normal forms of symplectic maps in R^4", Comp. Phys. Comm., **86**, pp. 199–207 (1995).

[9] W. Scandale, F. Schmidt, E. Todesco, "Compensation of the tune shift in the LHC, using the normal forms techniques", Part. Accel. **35**, pp. 53-81 (1991).

[10] M. Giovannozzi, R. Grassi, W. Scandale, E. Todesco, "Sorting approach to magnetic random errors", Phys. Rev. E **52**, pp. 3093–101 (1995).

[11] F. Schmidt and E. Todesco, "Evaluating high order resonances using resonant normal forms", in the proceedings of the the Fifth EPAC conference, Sitges, June 1996, http://wwwslap.cern.ch/frs/report/MOP034L.ps.Z.

[12] M. Giovannozzi et al., "PLATO: a program library for the analysis of nonlinear betatronic motion", CERN PS (PA) 96–12 (1996), submitted to Nuclear Instruments and Methods.

[13] E. Todesco, M. Gemmi and M. Giovannozzi, "NERO: a code for the Nonlinear Evaluation of Resonances in One-turn mappings", Comp. Phys. Commun., in press; documentation is also available at: /afs/cern.ch/project/lhcnap/nero.

[14] F. Schmidt, "SIXTRACK, version 1.2, single particle tracking code treating transverse motion with synchrotron oscillations in a symplectic manner", CERN SL/94–56 (AP) (1994), http://wwwslap.cern.ch/frs/Documentation/doc.html.

[15] M. Böge, H. Grote, Q. Qin and F. Schmidt, "Dynamic Aperture Studies for the LHC Version 4", LHC Project Report 31, in the proceedings of the Fifth EPAC Conference, Sitges, June 1996, http://wwwslap.cern.ch/frs/report/MOP033L.ps.Z.

[16] R. Bartolini and F. Schmidt, "Evaluation of Non-Linear Phase Space Distortions via Frequency Analysis", LHC Project Report 98 and in the proceedings of the workshop on: "Nonlinear and Collective Phenomena in Beam Physics", Arcidosso, September 1996, http://wwwslap.cern.ch/frs/report/arcif.ps.Z.

[17] Y. Cai and J.P. Koutchouk, "Correlations of dynamic aperture and resonance strengths for the LHC version 4.3 lattice", LHC Project Note 75.

[18] G. Guignard, "A general treatment of resonances in accelerators", CERN 78–11 (1978).

Neutralised beams: Landau Damping in System with Strong Nonlinearity

P.Zenkevich*, E.Mustafin*

*Institute for Theoretical and Experimental Physics
B.Cheremushkinskaya, 25, 117218, Moscow, Russia*

Abstract. It is well known that the electron current in a neutralised electron cooling system is limited by drift dipole instability which appears due to interaction between the cooling electron beam and the neutralising ions. This strong nonlinear system has specific features: there is only "internal" (for neutralising ions) nonlinearity due to ion Coulomb field inside the electron beam, and there is strong "external" nonlinearity outside the electron beam due to nonlinear field of the electrons. To investigate a mechanism of Landau damping in the system we calculated numerically free oscillations of the ion beam center of gravity for fixed electron beam center. The results have shown that "internal" nonlinearity does not lead to damping of this coherent oscillation, which may appear only due to "tail" ions interacting with "external" nonlinearity.

INTRODUCTION

In electron cooling systems such as LEAR ECOOL Ref. [1] it is important to use high density electron beam with minimal velocity spread. But large space charge of the beam causes large velocity spread of the electrons Refs. [2]- [3]. This space charge effects may be reduced by neutralisation of the electron beam with ions of the residual gas: electron beam ionizes the neutral residual gas atoms and molecules, these positive charged ions are captured and stored in the potential well of the electron beam up to some neutralisation degree depending on ion temperature, heating rate, pessure and other parameters of the system.

In 1992-1996 a possibility to improve the characteristics of LEAR ECOOL by neutralisation was examined experimentally and theoretically by collaboration CERN-Russia (JINR, ITEP) with participation of the authors of this paper. The investigations have confirmed that, probably, the main factor which limits the neutralisation degree and the current of the electron beam is the transverse dipole instability of the neutralised beam.

In order to describe the system of neutralised beam one may apply the "hydrodinamical" model Ref. [4], introducing into system some phenomenological damping constant. It is shown in this paper, that the theoretical approach of Ref. [4] (briefly described in the Section I) does not give a correct value of the damping constant. In order to estimate damping coefficient in particular case of the neutralised electron beam the special computer code has been developed. In the Section II the algorithm of the computer code is described. And in the Section III computed data are discussed.

I EQUATIONS OF MOTION

Let us consider beams with uniform transverse density. Moreover, we assume, that the electron beam is monochromatic. Then we may write:

$$\frac{dU_e}{dt} - i\omega_d^0 \sum_k \eta_k(U_e - U_i^k) = -i\frac{F_e(z)}{\omega_{Le}} \exp(-i\omega t), \quad (1)$$

and assuming the neutralising ions to be immovable in the longitudinal direction:

$$\frac{d^2 U_i^j}{dt^2} + i\omega_{Li}^j \frac{dU_i^j}{dt} + 2\gamma_\perp^j \frac{dU_i^j}{dt} + (\omega_{ion}^j)^2(U_i^j - U_e) - (\omega_{ion}^j)^2 \sum_k \eta_k(U_i^j - U_i^k)$$
$$= \frac{Z^j F_i(z)}{A^j} \exp(-i\omega t) \quad (2)$$

Here the subscripts "e" and "i" refer to the electrons and ions respectively, j, k refer to ion species, $\eta_k = Z^k n_k/n_e$ is the partial neutralization factor, $n_{e,k}$ is the density of the electrons and the ions respectively, $U = x + iy$ describes the horizontal (x) and vertical (y) displacement of the center of gravity of the beams in complex notation, t is a time. $F_{e,i}(z)\exp(-i\omega t)$ describe the external exciting fields. The different frequencies in Eqs.(1)-(2) are (cgs units, B in Gauss):

$\omega_{Le} = \frac{eB}{m_e c}$ Larmour frequency (=cyclotron frequency) of the electrons in the longitudinal magnetic field B

$\omega_{Li}^j = \frac{Z^j eB}{M^j c}$ Larmour frequency of the ions with charge number Z^j and mass M^j ($= A^j m_p$)

$\left(\omega_{ion}^j\right)^2 = \frac{2\pi n_e Z^j}{A^j} r_p c^2$ with ω_{ion}^j the frequency of the ion with number j in the space charge potential of the electrons, r_p is the classical proton radius

$\eta_j \omega_d^0 = \eta_j \frac{2\pi n_e ce}{B}$ the electron "drift frequency" in the space charge field of the species with number j

The phenomenological constant $\gamma_\perp^j$ describes the transverse Landau damping (LD). The derivative in Eq.(1) have to be taken along the orbit of the electrons i.e.

$$\frac{d}{dt} = \frac{\partial}{\partial t} + v_e \frac{\partial}{\partial z}, \qquad (3)$$

where v_e is the longitudinal velocity of the electrons.

Let us consider free oscillations without external exciting fields. If the neutralising ion beam consists of only one kind of ions (with definite ratio of A/Z), Eqs.(1)-(2) may be rewritten in the following form:

$$\begin{cases} v_e dV_e/dz - i\left[\omega + \omega_d^0 \eta\right] V_e + i\omega_d^0 \eta V_i = 0 \\ V_i = \omega_{ion}^2 V_e / \Delta_i \end{cases} \qquad (4)$$

where $U_e(z,t) = V_e(z)\exp(-i\omega t)$ describes the electron beam; $U_i(z,t) = V_i(z)\exp(-i\omega t)$ describes the standing ions; $\Delta_i = -\omega^2 + \omega_{Li}\omega + \omega_{ion}^2 - 2i\gamma_\perp \omega$.

It is easy to see from the last equations that the solution in the form of $V_{e,i}(z) = a_{e,i}\exp(ikz)$ leads to the wave number k, which at the coherent resonant frequency of the ions

$$\omega_{1,2}^{coh} = \frac{\omega_{Li}}{2} \pm \sqrt{\frac{\omega_{Li}^2}{4} + \omega_{ion}^2} \qquad (5)$$

has the imaginary part:

$$\text{Im}(k_{1,2}) = \eta \frac{\omega_d}{v_e} \frac{\omega_{ion}^2}{2\gamma_\perp \omega_{1,2}^{coh}} \qquad (6)$$

This solution describes the transverse drift instability with exponential growth of the coherent beam oscillation amplitude in space along the beam axis in the longitudinal direction. There is no absolute instability, i.e. growth of the amplitude in time.

The crucial point of the theory is how to determine damping constant $\gamma_\perp$. This linear model described above, of course, is very qualitative, since LD appears only due to nonlinearity of individual ion motion. Standard procedure to calculate the LD coefficient due to nonlinearity includes two steps: 1) solution of the stationary Vlasov's equation; 2) analysis of linearised Vlasov's equation for small pertrubations of the stationary solution. However, for two dimensional beams realisation of this procedure is obstacled by mathematical difficulties, and analytical estimation of the LD coefficient can be made only in the frame of some simplified models.

In Ref. [4] LD coefficient is derived by use of the following simplifying assumptions:

- the perturbed collective electric field is dipole, i.e. it is described by the following expression:
$$\mathbf{E} = \mathbf{E_0} \exp(-i\omega t) \qquad (7)$$
where $\mathbf{E_0}$ does not depend on space coordinates;
- the electron beam is almost completely compensated $(1 - \eta \ll 1)$ and its stationary distribution near the beam boundary is determined by Debye potential;
- the external field for ions (i.e. electric field of the electron beam) is linear;
- all the ions are inside the electron beam;
- ions are weakly magnetized $(\omega_{Li} \ll \omega_{ion})$:

In these assumptions LD appears due to nonlinearity of the ion stationary field. For $\gamma_\perp$ the following estimation has been obtained

$$\gamma_\perp = \sqrt{\frac{2T}{\pi M}} \frac{1}{a} \qquad (8)$$

where T is the ion temperature, a is the radius of the electron beam.

However, if one accepts third and fourth assumption then one immediately gets the well known solution for movement of the ion beam's center of gravity: there is no any damping in the system with linear external electric field. Indeed, let us write equation of motion for individual ion with number i:

$$\frac{d^2\mathbf{r_i}}{dt^2} + \omega_0^2 \mathbf{r_i} + \sum_j Ze\mathbf{E}(\mathbf{r_i} - \mathbf{r_j})/M = 0 \qquad (9)$$

where ω_0 corresponds to the linear external field, the function $\mathbf{E}(\mathbf{r_i} - \mathbf{r_j})$ is the electric field induced by the ion number j at the point with coordinates $\mathbf{r_i}$. Then for the center of gravity:

$$\bar{\mathbf{r}} = \sum \mathbf{r_i}$$

taking into account, that $\mathbf{E}(\mathbf{r_i} - \mathbf{r_j}) = -\mathbf{E}(\mathbf{r_j} - \mathbf{r_i})$, one gets the following equation:

$$\frac{d^2\bar{\mathbf{r}}}{dt^2} + \omega_0^2 \bar{\mathbf{r}} = 0 \qquad (10)$$

which corresponds to oscillation of the center of gravity of the ion beam without any damping.

We see that in this case $\gamma_\perp = 0$ and this result contradicts to Eq.(8). We belive that this contraiction appears because the first assumption is not correct. In such a system there is no pure dipole mode, and the perturbed field of oscillating beam depends on the transverse coordinates.

Thus correct value of damping coefficient is an open question and for its investigation in the case of the neutralised electron beam the special computer code has been developed.

II THE MODEL OF ION BEAM AND ALGORITHM OF THE COMPUTER CODE

Firstly, let us rewrite Eq.(2) for one ion specie without magnetic field (the case of weakly magnetized ions) and without the external field in the r.h.s.

$$\frac{d^2 U_i}{dt^2} + 2\gamma_\perp^j \frac{dU_i}{dt} + \omega_{ion}^2 (U_i - U_e) = 0 \tag{11}$$

Note, that if one fixes the electron beam center ($U_e = 0$) then the free oscillation of the ion beam damps with the decrement $\gamma_\perp$. Thus we may suggest following procedure to estimate the damping coefficient:

- the electron beam field may be considered as an external field for the ions:

$$E_{ext}(r) = 2\pi e n_e a \begin{cases} r/a & \text{for } r \leq a, \\ a/r & \text{for } r > a. \end{cases} \tag{12}$$

- the self-consistent ion distribution in the phase space should be "ergodic", i.e. it depends only on the Hamiltonian of the system (Ref. [5]). For numerical investigation it is convenient to use distribution of ions occupying a finite region in the phase space $\{x, y, p_x, p_y\}$ (2D problem in real space has been investigated). The case of uniform density has been used for computer simulation:

$$F(H) = \begin{cases} F_0, & H \leq H_0, \\ 0, & H > H_0. \end{cases} \tag{13}$$

with H_0 is the ion maximal energy. Note that the ion beam radius b depends on the maximal energy H_0 and on the neutralisation degree as well. The ions with the energy large enough to go outside the electron beam will be influenced by strongly nonlinear (even decreasing with the radius) electric field. The case of $b/a = 1.5$ has been used for numerical simulation.

- after preparing the self-consistent distribution of ions Eq.(13) in the external field of Eq.(12) (using formulae derived in Ref. [5]) one may give a shift to the ion beam as a whole from the center of the external field and look at the evolution of the ion beam center of gravity in time. If there is any damping of ion beam center of gravity oscillations due to nonlinearity of the electric field then the decrement may be estimated.

The PM (Particle-Mesh) Method of computer simulation and following methods for subroutines have been applied (see Ref. [6]):

- "leapfrog" method to integrate the equations of motion

- "cloud-in-cell" method to distribute charge of the macroparticles into cells, with a "cloud" of rectangular shape

- the method of Fast Fourrier Transformation to integrate Poisson equation

The typical number of cells used in the calculations is 129x129 or 257x257, the number of "macroparticles" is $5 \bullet 10^5 \div 8 \bullet 10^5$.

III RESULTS OF SIMULATION

In order to confirm validity of the developed computer code numerical solutions have been compared with known analytical ones. If instead of Eq.(13) one choose the δ-function distribution in the space of Hamiltonian H, and assume that the ion beam size coincides to the electron beam size, which means, that the external field for the ions is purely linear, then one gets the problem analytically investigated in Ref. [7].

The computer simulations show that for this case the ion beam center of gravity oscillations are purely sinusoidal (there are no damping). Its frequency does not depend on the neutralisation degree and depends only on the external field which stays in agreement with results of Eq.(10) and Ref. [7]

According to Ref. [7] two quadrupole modes of oscillations may be distinguished, which linearly depend on the neutralisation degree. Fig.(1) shows the good agreement of analytical solution with the results of simulation.

After validation the code has been applied to investigate the model of the ion beam described in Section II.

Fixing the relation $b/a = 1.5$ and varying the neutralisation degree η evolution of the ion beam center of gravity has been calculated. The typical behavior of oscillations for the cases of $\eta = 0.2$ and $\eta = 0.8$ is shown in Figs.(2)-(3). The Fourrier-spectrum for $\eta = 0.2$ and $\eta = 0.8$ are plotted in Fig.(4) and Fig.(5) One may distinguish oscillations of at least two modes in Fig.(2). One of them has wide band and corresponds to the fast damping mode, another one corresponds to the mode which does not damp or damps very weakly (beside the precision of the code). These two modes have been distinguished for values of neutralisation degree in the range of $\eta = 0.1 \div 0.9$, although for higher values of η the amplitude of the fast damping mode is small compare to the nondamping mode.

The dependence of frequency of nondamping mode corresponding to the higher narrow peaks in Fig.(3) and Fig.(4) on the neutralisation degree is shown in Fig.(6).

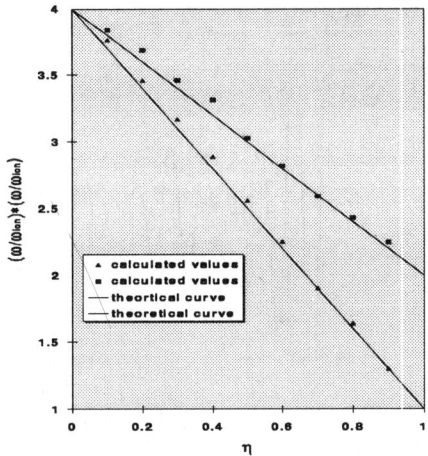

FIGURE 1. Test calculations: Dependence of quadrupole mode frequency on η.

IV DISCUSSION

Analysis of free oscillations of the ion beam center of gravity does not confirm the model developed in Ref. [4]. It is shown that "self-stabilization" of these oscillations due to nonlinearity of the ion beam Coulomb field is absent. This result coincides with "one-mode" theory for sheet beams Ref. [8], which shows that "self-stabilization" exists only for modes with higher betatron number n ($n \geq 3$). From the point of view of the standard accelerator theory this effect is connected with well known fact that for mode with $n = 1$ the coherent frequency shift is much more that the frequency spread due to this internal nonlinearity.

However, the presence of particles in the ion beam "tail" (outside the electron beam) makes the result more complicated and modes with fast damping appear. Amplitude of these modes depend on b/a and netralisation degree (it is clear that this damping should increase with b/a and decrease with η).

These results should be considered as preliminary ones since to check its validity it is necessary to perform 3D simulation of the coupled electron-ion oscillations which is outside of our present calculation possibilities.

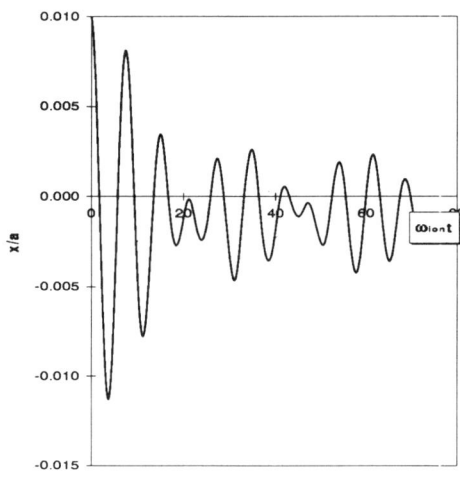

FIGURE 2. Oscillation of the ion beam center of gravity, $\eta = 0.2$.

ACKNOWLEDGEMENTS

The authors are very grateful to Dr. D.Moehl (CERN) who suggested us to paticipate in CERN-Russia collaboration on neutralised beam investigations, and to Dr. Z.Parsa for invitation to take part in this very interesting seminar.

REFERENCES

1. J.Bosser, *Electron Cooling*, CERN/PS 92-18 (BD), 1992
2. J.Bosser, I.Meshkov et al *Project for a Variable Current Electron Gun for the LEAR Electron Cooler*, CERN/PS 92-03 (AR), 1992
3. J.Bosser, I.Meshkov, V.Parkhomchuk et al *Neutralisation of the LEAR-ECOOL Electron Beam Space Charge* CERN/PS 93-08 (AR), 1993
4. A.V.Burov, V.I. Kudelainen et al, *Experimental Investigation of an Electron Beam in Compensated State*, CERN/PS 93-03 (AR).
5. I.Meshkov, E.Syresin, E.Mustafin, P.Zenkevich, "Stationary Parameters of Neutralised Electron Cooling System", in *Proceedings of the Conference "Beam Cooling and Damping"*, June, 1996, Dubna, forthcoming.
6. R.W.Hockney, J.W.Eastwood *Computer Simulation Using Particles*, 1981, McGraw-Hill Inc.

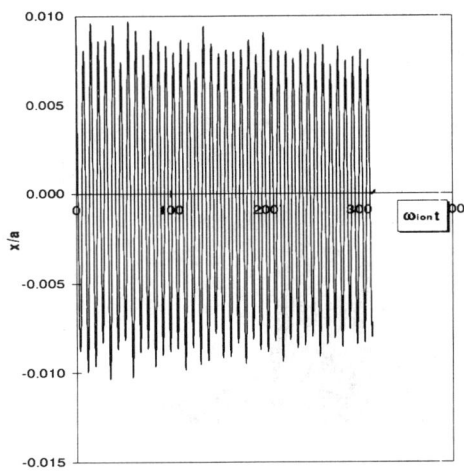

FIGURE 3. Oscillation of the ion beam center of gravity, $\eta = 0.8$.

7. R.L.Gluckstern "Oscillation Modes in Two Dimensional Beams", in *Proceedings of the Linear Accelerator Conference 1970*, FNAL, Batavia, 1970, Vol. 2, pp. 811-819.
8. P.Zenkevich, A.Korolev *Self-Stabilization of Coherent Oscillations of Sheet Beams*, Moscow, 1985, Preprint ITEP-173.

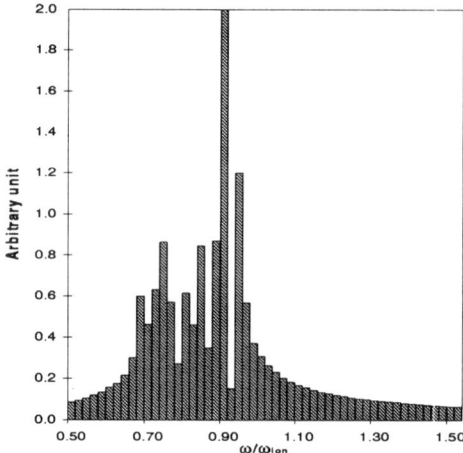

FIGURE 4. Spectrum of ion beam center of gravity oscillation, $\eta = 0.2$.

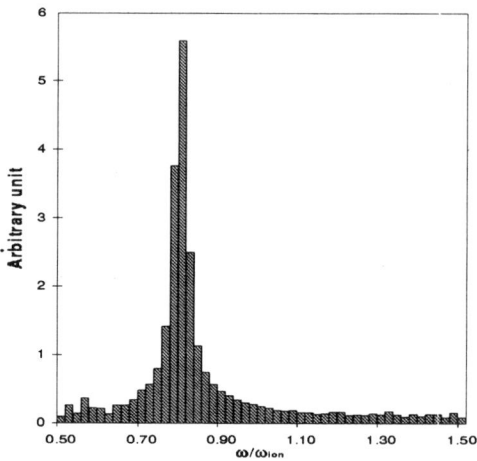

FIGURE 5. Spectrum of ion beam center of gravity oscillation, $\eta = 0.8$.

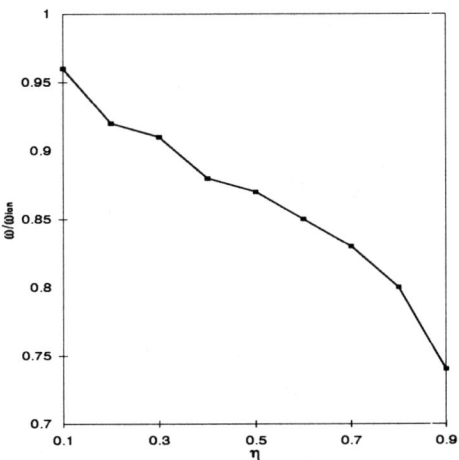

FIGURE 6. Dependence of nondamping dipole mode frequency on η.

Nonlinear Dynamics of Accelerator via Wavelet Approach

A.N. Fedorova and M.G. Zeitlin

Institute of Problems of Mechanical Engineering,
Russian Academy of Sciences, Russia, 199178, St. Petersburg,
V.O., Bolshoj pr., 61, e-mail: zeitlin@math.ipme.ru

Abstract. In this paper we present the applications of methods from wavelet analysis to polynomial approximations for a number of accelerator physics problems. In the general case we have the solution as a multiresolution expansion in the base of compactly supported wavelet basis. The solution is parametrized by the solutions of two reduced algebraical problems, one is nonlinear and the second is some linear problem, which is obtained from one of the next wavelet constructions: Fast Wavelet Transform, Stationary Subdivision Schemes, the method of Connection Coefficients. According to the orbit method and by using construction from the geometric quantization theory we construct the symplectic and Poisson structures associated with generalized wavelets by using metaplectic structure. We consider wavelet approach to the calculations of Melnikov functions in the theory of homoclinic chaos in perturbed Hamiltonian systems and for parametrization of Arnold–Weinstein curves in Floer variational approach.

Introduction. In this paper we consider the following problems: the calculation of orbital motion in storage rings, some aspects of symplectic invariant approach to wavelet computations, Melnikov functions approach in the theory of homoclinic chaos, the calculation of Arnold-Weinstein curves (periodic loops) in Hamiltonian systems. The key point in the solution of these problems is the use of the methods of wavelet analysis, relatively novel set of mathematical methods, which gives us a possibility to work with well-localized bases in functional spaces and with the general type of operators (including pseudodifferential) in such bases. Our problem as many related problems in the framework of our type of approximations of complicated physical nonlinearities is reduced to the problem of the solving of the systems of differential equations with polynomial nonlinearities with or without some constraints. In this paper we consider as the main example the particle motion in storage rings in standard approach, which is based on consideration in [1], [2]. Starting from Hamiltonian, which described classical dynamics in storage rings

$$\mathcal{H}(\vec{r}, \vec{P}, t) = c\{\pi^2 + m_0^2 c^2\}^{1/2} + e\phi \qquad (1)$$

and using Serret–Frenet parametrization, we have the following Hamiltonian for orbital motion in machine coordinates:

$$\mathcal{H}(x, p_x, z, p_z, \sigma, p_\sigma; s) = p_\sigma - [1 + f(p_\sigma)] \cdot [1 + K_x \cdot x + K_z \cdot z] \times \qquad (2)$$

$$\left\{ 1 - \frac{[p_x + H \cdot z]^2 + [p_z - H \cdot x]^2}{[1 + f(p_\sigma)]^2} \right\}^{1/2}$$

$$+ \frac{1}{2} \cdot [1 + K_x \cdot x + K_z \cdot z]^2 - \frac{1}{2} \cdot g \cdot (z^2 - x^2) - N \cdot xz$$

$$+ \frac{\lambda}{6} \cdot (x^3 - 3xz^2) + \frac{\mu}{24} \cdot (z^4 - 6x^2 z^2 + x^4)$$

$$+ \frac{1}{\beta_0^2} \cdot \frac{L}{2\pi \cdot h} \cdot \frac{eV(s)}{E_0} \cdot \cos\left[h \cdot \frac{2\pi}{L} \cdot \sigma + \varphi\right]$$

Then, after standard manipulations with truncation of power series expansion of square root we arrive to the following approximated Hamiltonian for particle motion:

$$\mathcal{H} = \frac{1}{2} \cdot \frac{[p_x + H \cdot z]^2 + [p_z - H \cdot x]^2}{[1 + f(p_\sigma)]} + p_\sigma - [1 + K_x \cdot x + K_z \cdot z] \qquad (3)$$

$$\cdot f(p_\sigma) + \frac{1}{2} \cdot [K_x^2 + g] \cdot x^2 + \frac{1}{2} \cdot [K_z^2 - g] \cdot z^2 - N \cdot xz +$$

$$\frac{\lambda}{6} \cdot (x^3 - 3xz^2) + \frac{\mu}{24} \cdot (z^4 - 6x^2 z^2 + x^4)$$

$$+ \frac{1}{\beta_0^2} \cdot \frac{L}{2\pi \cdot h} \cdot \frac{eV(s)}{E_0} \cdot \cos\left[h \cdot \frac{2\pi}{L} \cdot \sigma + \varphi\right]$$

and the corresponding equations of motion:

$$\frac{d}{ds} x = \frac{\partial \mathcal{H}}{\partial p_x} = \frac{p_x + H \cdot z}{[1 + f(p_\sigma)]};$$

$$\frac{d}{ds} p_x = -\frac{\partial \mathcal{H}}{\partial x} = \frac{[p_z - H \cdot x]}{[1 + f(p_\sigma)]} \cdot H - [K_x^2 + g] \cdot x + N \cdot z +$$

$$K_x \cdot f(p_\sigma) - \frac{\lambda}{2} \cdot (x^2 - z^2) - \frac{\mu}{6}(x^3 - 3xz^2); \qquad (4)$$

$$\frac{d}{ds} z = \frac{\partial \mathcal{H}}{\partial p_z} = \frac{p_z - H \cdot x}{[1 + f(p_\sigma)]};$$

$$\frac{d}{ds} p_z = -\frac{\partial \mathcal{H}}{\partial z} = -\frac{[p_x + H \cdot z]}{[1 + f(p_\sigma)]} \cdot H - [K_z^2 - g] \cdot z + N \cdot x +$$

$$K_z \cdot f(p_\sigma) - \lambda \cdot xz - \frac{\mu}{6}(z^3 - 3x^2 z);$$

$$\frac{d}{ds}\sigma = \frac{\partial \mathcal{H}}{\partial p_\sigma} = 1 - [1 + K_x \cdot x + K_z \cdot z] \cdot f'(p_\sigma) -$$
$$\frac{1}{2} \cdot \frac{[p_x + H \cdot z]^2 + [p_z - H \cdot x]^2}{[1 + f(p_\sigma)]^2} \cdot f'(p_\sigma)$$
$$\frac{d}{ds}p_\sigma = -\frac{\partial \mathcal{H}}{\partial \sigma} = \frac{1}{\beta_0^2} \cdot \frac{eV(s)}{E_0} \cdot \sin\left[h \cdot \frac{2\pi}{L} \cdot \sigma + \varphi\right]$$

Then we use series expansion of function $f(p_\sigma)$ from [2]:

$$f(p_\sigma) = f(0) + f'(0)p_\sigma + f''(0)\frac{1}{2}p_\sigma^2 + \ldots = p_\sigma - \frac{1}{\gamma_0^2} \cdot \frac{1}{2}p_\sigma^2 + \ldots$$

and the corresponding expansion of RHS of equations (4). In the following we take into account only an arbitrary polynomial (in terms of dynamical variables) expressions and neglecting all nonpolynomial types of expressions, i.e. we consider such approximations of RHS, which are not more than polynomial functions in dynamical variables and arbitrary functions of independent variable s ("time" in our case, if we consider our system of equations as dynamical problem).

I POLYNOMIAL DYNAMICS

Introduction. The first main part of our consideration is some variational approach to this problem, which reduces initial problem to the problem of solution of functional equations at the first stage and some algebraical problems at the second stage. We consider also two private cases of our general construction. In the first case (particular) we have for Riccati equations (particular quadratic approximations) the solution as a series on shifted Legendre polynomials, which is parameterized by the solution of reduced algebraical (also Riccati) system of equations. This is only an example of general construction. In the second case (general polynomial system) we have the solution in a compactly supported wavelet basis. Multiresolution expansion is the second main part of our construction. The solution is parameterized by solutions of two reduced algebraical problems, one as in the first case and the second is some linear problem, which is obtained from one of the next wavelet construction: Fast Wavelet Transform (FWT), Stationary Subdivision Schemes (SSS), the method of Connection Coefficients (CC).

Variational method. Our problems may be formulated as the systems of ordinary differential equations $dx_i/dt = f_i(x_j, t)$, $(i, j = 1, \ldots, n)$ with fixed initial conditions $x_i(0)$, where f_i are not more than polynomial functions of dynamical variables x_j and have arbitrary dependence of time. Because of time dilation we can consider only next time interval: $0 \leq t \leq 1$. Let us consider a set of functions $\Phi_i(t) = x_i dy_i/dt + f_i y_i$ and a set of functionals

$F_i(x) = \int_0^1 \Phi_i(t)dt - \overline{x_i y_i}\,|_0^1$, where $y_i(t)(y_i(0) = 0)$ are dual variables. It is obvious that the initial system and the system $F_i(x) = 0$ are equivalent. In the last part we consider the symplectization of this approach. Now we consider formal expansions for x_i, y_i:

$$x_i(t) = x_i(0) + \sum_k \lambda_i^k \varphi_k(t) \quad y_j(t) = \sum_r \eta_j^r \varphi_r(t), \tag{5}$$

where because of initial conditions we need only $\varphi_k(0) = 0$. Then we have the following reduced algebraical system of equations on the set of unknown coefficients λ_i^k of expansions (5):

$$\sum_k \mu_{kr} \lambda_i^k - \gamma_i^r(\lambda_j) = 0 \tag{6}$$

Its coefficients are $\mu_{kr} = \int_0^1 \varphi_k'(t)\varphi_r(t)dt$, $\gamma_i^r = \int_0^1 f_i(x_j, t)\varphi_r(t)dt$. Now, when we solve system (6) and determine unknown coefficients from formal expansion (5) we therefore obtain the solution of our initial problem. It should be noted if we consider only truncated expansion (5) with N terms then we have from (6) the system of $N \times n$ algebraical equations and the degree of this algebraical system coincides with degree of initial differential system. So, we have the solution of the initial nonlinear (polynomial) problem in the form

$$x_i(t) = x_i(0) + \sum_{k=1}^N \lambda_i^k X_k(t), \tag{7}$$

where coefficients λ_i^k are roots of the corresponding reduced algebraical problem (6). Consequently, we have a parametrization of solution of initial problem by solution of reduced algebraical problem (6). But in general case, when the problem of computations of coefficients of reduced algebraical system (6) is not solved explicitly as in the quadratic case, which we shall consider below, we have also parametrization of solution (4) by solution of corresponding problems, which appear when we need to calculate coefficients of (6). As we shall see, these problems may be explicitly solved in wavelet approach.

The solutions Next we consider the construction of explicit time solution for our problem. The obtained solutions are given in the form (7), where in our first case we have $X_k(t) = Q_k(t)$, where $Q_k(t)$ are shifted Legendre polynomials and λ_k^i are roots of reduced quadratic system of equations. In wavelet case $X_k(t)$ correspond to multiresolution expansions in the base of compactly supported wavelets and λ_k^i are the roots of corresponding general polynomial system (6) with coefficients, which are given by FWT, SSS or CC constructions. According to the variational method to give the reduction from differential to algebraical system of equations we need compute the objects γ_a^j and μ_{ji}, which are constructed from objects:

$$\sigma_i \equiv \int_0^1 X_i(\tau)d\tau = (-1)^{i+1}, \quad \nu_{ij} \equiv \int_0^1 X_i(\tau)X_j(\tau)d\tau = \sigma_i\sigma_j + \frac{\delta_{ij}}{(2j+1)},$$

$$\mu_{ji} \equiv \int X_i'(\tau)X_j(\tau)d\tau = \sigma_j F_1(i,0) + F_1(i,j), \tag{8}$$

$$F_1(r,s) = [1 - (-1)^{r+s}]\hat{s}(r-s-1), \quad \hat{s}(p) = \begin{cases} 1, & p \geq 0 \\ 0, & p < 0 \end{cases}$$

$$\beta_{klj} \equiv \int_0^1 X_k(\tau)X_l(\tau)X_j(\tau)d\tau = \sigma_k\sigma_l\sigma_j +$$

$$\alpha_{klj} + \frac{\sigma_k \delta_{jl}}{2j+1} + \frac{\sigma_l \delta_{kj}}{2k+1} + \frac{\sigma_j \delta_{kl}}{2l+1},$$

$$\alpha_{klj} \equiv \int_0^1 X_k^* X_l^* X_j^* d\tau = \frac{1}{(j+k+l+1)R(1/2(i+j+k))} \times$$
$$R(1/2(j+k-l))R(1/2(j-k+l))R(1/2(-j+k+l)),$$

if $j + k + l = 2m, m \in Z$, and $\alpha_{klj} = 0$ if $j + k + l = 2m + 1$; where $R(i) = (2i)!/(2^i i!)^2$, $Q_i = \sigma_i + P_i^*$, where the second equality in the formulae for $\sigma, \nu, \mu, \beta, \alpha$ hold for the first case.

Wavelet computations. Now we give construction for computations of objects (8) in the wavelet case. We use some constructions from multiresolution analysis: a sequence of successive approximation closed subspaces V_j: ...$V_2 \subset V_1 \subset V_0 \subset V_{-1} \subset V_{-2} \subset$ ... satisfying the following properties: $\bigcap_{j \in \mathbf{Z}} V_j = 0$, $\overline{\bigcup_{j \in \mathbf{Z}} V_j} = L^2(\mathbf{R})$, $f(x) \in V_j <=> f(2x) \in V_{j+1}$ There is a function $\varphi \in V_0$ such that $\{\varphi_{0,k}(x) = \varphi(x-k)_{k \in \mathbf{Z}}\}$ forms a Riesz basis for V_0. We use compactly supported wavelet basis: orthonormal basis for functions in $L^2(\mathbf{R})$. As usually $\varphi(x)$ is a scaling function, $\psi(x)$ is a wavelet function, where $\varphi_i(x) = \varphi(x - i)$. Scaling relation that defines φ, ψ are

$$\varphi(x) = \sum_{k=0}^{N-1} a_k \varphi(2x-k) = \sum_{k=0}^{N-1} a_k \varphi_k(2x), \quad \psi(x) = \sum_{k=-1}^{N-2} (-1)^k a_{k+1} \varphi(2x+k)$$

Let be $f : \mathbf{R} \longrightarrow \mathbf{C}$ and the wavelet expansion is

$$f(x) = \sum_{\ell \in \mathbf{Z}} c_\ell \varphi_\ell(x) + \sum_{j=0}^{\infty} \sum_{k \in \mathbf{Z}} c_{jk} \psi_{jk}(x) \tag{9}$$

The indices k, ℓ and j represent translation and scaling, respectively

$$\varphi_{jl}(x) = 2^{j/2} \varphi(2^j x - \ell), \psi_{jk}(x) = 2^{j/2} \psi(2^j x - k)$$

The set $\{\varphi_{j,k}\}_{k \in \mathbf{Z}}$ forms a Riesz basis for V_j. Let W_j be the orthonormal complement of V_j with respect to V_{j+1}. Just as V_j is spanned by dilation and translations of the scaling function, so are W_j spanned by translations and dilation of the mother wavelet $\psi_{jk}(x)$. If in formulae (9) $c_{jk} = 0$ for $j \geq J$, then $f(x)$ has an alternative expansion in terms of dilated scaling functions only $f(x) = \sum_{\ell \in \mathbb{Z}} c_{J\ell} \varphi_{J\ell}(x)$. This is a finite wavelet expansion, it can be written solely in terms of translated scaling functions. We use wavelet $\psi(x)$, which has k vanishing moments $\int x^k \psi(x) d(x) = 0$, or equivalently $x^k = \sum c_\ell \varphi_\ell(x)$ for each k, $0 \leq k \leq K$. Also we have the shortest possible support: scaling function DN (where N is even integer) will have support $[0, N-1]$ and $N/2$ vanishing moments. There exists $\lambda > 0$ such that DN has λN continuous derivatives; for small $N, \lambda \geq 0.55$. To solve our second associated linear problem we need to evaluate derivatives of $f(x)$ in terms of $\varphi(x)$. Let be $\varphi_\ell^n = d^n \varphi_\ell(x)/dx^n$. We derive the wavelet - Galerkin approximation of a differentiated $f(x)$ as $f^d(x) = \sum_\ell c_\ell \varphi_\ell^d(x)$ and values $\varphi_\ell^d(x)$ can be expanded in terms of $\varphi(x)$

$$\phi_\ell^d(x) = \sum_m \lambda_m \varphi_m(x), \quad \lambda_m = \int_{-\infty}^{\infty} \varphi_\ell^d(x) \varphi_m(x) dx$$

The coefficients λ_m are 2-term connection coefficients. In general we need to find ($d_i \geq 0$)

$$\Lambda_{\ell_1 \ell_2 ... \ell_n}^{d_1 d_2 ... d_n} = \int_{-\infty}^{\infty} \prod \varphi_{\ell_i}^{d_i}(x) dx \qquad (10)$$

For Riccati case we need to evaluate two and three connection coefficients

$$\Lambda_\ell^{d_1 d_2} = \int_{-\infty}^{\infty} \varphi^{d_1}(x) \varphi_\ell^{d_2}(x) dx, \quad \Lambda^{d_1 d_2 d_3} = \int_{-\infty}^{\infty} \varphi^{d_1}(x) \varphi_\ell^{d_2}(x) \varphi_m^{d_3}(x) dx$$

According to CC method [12] we use the next construction. When N in scaling equation is a finite even positive integer the function $\varphi(x)$ has compact support contained in $[0, N-1]$. For a fixed triple (d_1, d_2, d_3) only some $\Lambda_{\ell m}^{d_1 d_2 d_3}$ are nonzero : $2 - N \leq \ell \leq N - 2$, $2 - N \leq m \leq N - 2$, $|\ell - m| \leq N - 2$. There are $M = 3N^2 - 9N + 7$ such pairs (ℓ, m). Let $\Lambda^{d_1 d_2 d_3}$ be an M-vector, whose components are numbers $\Lambda_{\ell m}^{d_1 d_2 d_3}$. Then we have the first key result: Λ satisfy the system of equations ($d = d_1 + d_2 + d_3$)

$$A \Lambda^{d_1 d_2 d_3} = 2^{1-d} \Lambda^{d_1 d_2 d_3}, \quad A_{\ell, m; q, r} = \sum_p a_p a_{q - 2\ell + p} a_{r - 2m + p}$$

By moment equations we have created a system of $M + d + 1$ equations in M unknowns. It has rank M and we can obtain unique solution by combination of LU decomposition and QR algorithm. The second key result gives us the 2-term connection coefficients:

$$A\Lambda^{d_1 d_2} = 2^{1-d}\Lambda^{d_1 d_2}, \quad d = d_1 + d_2, \quad A_{\ell,q} = \sum_p a_p a_{q-2\ell+p}$$

For nonquadratic case we have analogously additional linear problems for objects (10). Also, we use FWT and SSS for computing coefficients of reduced algebraic systems. We use for modelling D6,D8,D10 functions and programs RADAU and DOPRI for testing.

As a result we obtained the explicit time solution (7) of our problem. In comparison with wavelet expansion on the real line which we use now and in calculation of Galerkin approximation, Melnikov function approach, etc also we need to use periodized wavelet expansion, i.e. wavelet expansion on finite interval. Also in the solution of perturbed system we have some problem with variable coefficients. For solving last problem we need to consider one more refinement equation for scaling function $\phi_2(x)$: $\phi_2(x) = \sum_{k=0}^{N-1} a_k^2 \phi_2(2x - k)$ and corresponding wavelet expansion for variable coefficients $b(t)$: $\sum_k B_k^j(b)\phi_2(2^j x - k)$, where $B_k^j(b)$ are functionals supported in a small neighborhood of $2^{-j}k$.

The solution of the first problem consists in periodizing. In this case we use expansion into periodized wavelets defined by $\phi^{per}_{-j,k}(x) = 2^{j/2}\sum_Z \phi(2^j x + 2^j \ell - k)$. All these modifications lead only to transformations of coefficients of reduced algebraic system, but general scheme remains the same.

II METAPLECTIC WAVELETS

In this part we continue the application of powerful methods of wavelet analysis to polynomial approximations of nonlinear accelerator physics problems. In part 1 we considered our main example and general approach for constructing wavelet representation for orbital motion in storage rings. But now we need take into account the Hamiltonian or symplectic structure related with system (4). Therefore, we need to consider generalized wavelets, which allow us to consider the corresponding symplectic structures, instead of compactly supported wavelet representation. By using the orbit method and constructions from the geometric quantization theory we consider the symplectic and Poisson structures associated with Weyl- Heisenberg wavelets by using metaplectic structure and the corresponding polarization. In the next part we consider applications to construction of Melnikov functions in the theory of homoclinic chaos in perturbed Hamiltonian systems.

In wavelet analysis the following three concepts are used now: 1). a square integrable representation U of a group G, 2). coherent states over G, 3). the wavelet transform associated to U.

We have three important particular cases:

a) the affine $(ax+b)$ group, which yields the usual wavelet analysis

$$[\pi(b,a)f](x) = \frac{1}{\sqrt{a}} f\left(\frac{x-b}{a}\right)$$

b). the Weyl-Heisenberg group which leads to the Gabor functions, i.e. coherent states associated with windowed Fourier transform.

$$[\pi(q,p,\varphi)f](x) = \exp(i\mu(\varphi - p(x-q)))f(x-q)$$

In both cases time-frequency plane corresponds to the phase space of group representation.

c). also, we have the case of bigger group, containing both affine and Weyl-Heisenberg group, which interpolate between affine wavelet analysis and windowed Fourier analysis: affine Weyl–Heisenberg group [13]. But usual representation of it is not square–integrable and must be modified: restriction of the representation to a suitable quotient space of the group (the associated phase space in that case) restores square – integrability: $G_{aWH} \longrightarrow$ homogeneous space. Also, we have more general approach which allows to consider wavelets corresponding to more general groups and representations [14], [15]. Our goal is applications of these results to problems of Hamiltonian dynamics and as consequence we need to take into account symplectic nature of our dynamical problem. Also, the symplectic and wavelet structures must be consistent (this must be resemble the symplectic or Lie-Poisson integrator theory). We use the point of view of geometric quantization theory (orbit method) instead of harmonic analysis. Because of this we can consider (a) – (c) analogously.

Metaplectic Group and Representations. Let $Sp(n)$ be symplectic group, $Mp(n)$ be its unique two- fold covering – metaplectic group. Let V be a symplectic vector space with symplectic form (,), then $R \oplus V$ is nilpotent Lie algebra - Heisenberg algebra:

$$[R,V] = 0, \quad [v,w] = (v,w) \in R, \quad [V,V] = R.$$

$Sp(V)$ is a group of automorphisms of Heisenberg algebra.

Let N be a group with Lie algebra $R \oplus V$, i.e. Heisenberg group. By Stone– von Neumann theorem Heisenberg group has unique irreducible unitary representation in which $1 \mapsto i$. This representation is projective: $U_{g_1} U_{g_2} = c(g_1,g_2) \cdot U_{g_1 g_2}$, where c is a map: $Sp(V) \times Sp(V) \to S^1$, i.e. c is S^1-cocycle.

But this representation is unitary representation of universal covering, i.e. metaplectic group $Mp(V)$. We give this representation without Stone-von Neumann theorem. Consider a new group $F = N' \bowtie Mp(V)$, $\bowtie$ is semidirect product (we consider instead of $N = R \oplus V$ the $N' = S^1 \times V$, $S^1 = (R/2\pi Z)$). Let V^* be dual to V, $G(V^*)$ be automorphism group of V^*. Then F is subgroup of $G(V^*)$, which consists of elements, which acts on V^* by affine transformations.

This is the key point!

Let $q_1, ..., q_n; p_1, ..., p_n$ be symplectic basis in V, $\alpha = pdq = \sum p_i dq_i$ and $d\alpha$ be symplectic form on V^*. Let M be fixed affine polarization, then for $a \in F$ the map $a \mapsto \Theta_a$ gives unitary representation of G: $\Theta_a : H(M) \to H(M)$

Explicitly we have for representation of N on H(M):

$$(\Theta_q f)^*(x) = e^{-iqx} f(x), \quad \Theta_p f(x) = f(x - p)$$

The representation of N on H(M) is irreducible. Let A_q, A_p be infinitesimal operators of this representation

$$A_q = \lim_{t \to 0} \frac{1}{t}[\Theta_{-tq} - I], \quad A_p = \lim_{t \to 0} \frac{1}{t}[\Theta_{-tp} - I],$$

then $\quad A_q f(x) = i(qx)f(x), \quad A_p f(x) = \sum p_j \frac{\partial f}{\partial x_j}(x)$

Now we give the representation of infinitesimal basic elements. Lie algebra of the group F is the algebra of all (nonhomogeneous) quadratic polynomials of (p,q) relatively Poisson bracket (PB). The basis of this algebra consists of elements $1, q_1, ..., q_n, p_1, ..., p_n, q_i q_j, q_i p_j, p_i p_j$, $i, j = 1, ..., n$, $i \leq j$,

PB is $\{f, g\} = \sum \frac{\partial f}{\partial p_j} \frac{\partial g}{\partial q_i} - \frac{\partial f}{\partial q_i} \frac{\partial g}{\partial p_i}$ and $\{1, g\} = 0$ for all g,

$\{p_i, q_j\} = \delta_{ij}$, $\{p_i q_j, q_k\} = \delta_{ik} q_j$, $\{p_i q_j, p_k\} = -\delta_{jk} p_i$, $\{p_i p_j, p_k\} = 0$,
$\{p_i p_j, q_k\} = \delta_{ik} p_j + \delta_{jk} p_i$, $\{q_i q_j, q_k\} = 0$, $\{q_i q_j, p_k\} = -\delta_{ik} q_j - \delta_{jk} q_i$

so, we have the representation of basic elements $f \mapsto A_f : 1 \mapsto i, q_k \mapsto ix_k$,

$$p_l \mapsto \frac{\delta}{\delta x^l}, p_i q_j \mapsto x^i \frac{\partial}{\partial x^j} + \frac{1}{2} \delta_{ij}, \quad p_k p_l \mapsto \frac{1}{i} \frac{\partial^k}{\partial x^k \partial x^l}, q_k q_l \mapsto ix^k x^l$$

This gives the structure of the Poisson manifolds to representation of any (nilpotent) algebra or in other words to continuous wavelet transform.

The Segal-Bargman Representation. Let $z = 1/\sqrt{2} \cdot (p - iq)$, $\bar{z} = 1/\sqrt{2} \cdot (p + iq)$, $p = (p_1, ..., p_n)$, F_n is the space of holomorphic functions of n complex variables with $(f, f) < \infty$, where

$$(f, g) = (2\pi)^{-n} \int f(z) \overline{g(z)} e^{-|z|^2} dp dq$$

Consider a map $U : H \to F_n$, where H is with real polarization, F_n is with complex polarization, then we have

$$(U\Psi)(a) = \int A(a, q) \Psi(q) dq, \quad \text{where} \quad A(a, q) = \pi^{-n/4} e^{-1/2(a^2 + q^2) + \sqrt{2} aq}$$

i.e. the Bargmann formula produce wavelets. We also have the representation of Heisenberg algebra on F_n :

$$U\frac{\partial}{\partial q_j}U^{-1} = \frac{1}{\sqrt{2}}\left(z_j - \frac{\partial}{\partial z_j}\right), \quad Uq_jU^{-1} = -\frac{i}{\sqrt{2}}\left(z_j + \frac{\partial}{\partial z_j}\right)$$

and also : $\omega = d\beta = dp \wedge dq$, where $\beta = i\bar{z}dz$.

Orbital Theory for Wavelets. Let coadjoint action be $< g \cdot f, Y > = < f, Ad(g)^{-1}Y >$, where $<,>$ is pairing $g \in G$, $f \in g^*$, $Y \in \mathcal{G}$. The orbit is $\mathcal{O}_f = G \cdot f \equiv G/G(f)$. Also, let A=A(M) be algebra of functions, V(M) is A-module of vector fields, A^p is A-module of p-forms. Vector fields on orbit is

$$\sigma(\mathcal{O}, X)_f(\phi) = \frac{d}{dt}(\phi(\exp tXf))\Big|_{t=0}$$

where $\phi \in A(\mathcal{O})$, $f \in \mathcal{O}$. Then $\mathcal{O}_f$ are homogeneous symplectic manifolds with 2-form $\Omega(\sigma(\mathcal{O}, X)_f, \sigma(\mathcal{O}, Y)_f) = < f, [X, Y] >$, and $d\Omega = 0$. PB on $\mathcal{O}$ have the next form $\{\Psi_1, \Psi_2\} = p(\Psi_1)\Psi_2$ where p is $A^1(\mathcal{O}) \to V(\mathcal{O})$ with definition $\Omega(p(\alpha), X) = i(X)\alpha$. Here $\Psi_1, \Psi_2 \in A(\mathcal{O})$ and $A(\mathcal{O})$ is Lie algebra with bracket $\{,\}$. Now let N be a Heisenberg group. Consider adjoint and coadjoint representations in some particular case. $N = (z, t) \in C \times R, z = p + iq$; compositions in N are $(z, t) \cdot (z', t') = (z + z', t + t' + B(z, z'))$, where $B(z, z') = pq' - qp'$. Inverse element is $(-t, -z)$. Lie algebra n of N is $(\zeta, \tau) \in C \times R$ with bracket $[(\zeta, \tau), (\zeta', \tau')] = (0, B(\zeta, \zeta'))$. Centre is $\tilde{z} \in n$ and generated by $(0,1)$; Z is a subgroup $\exp \tilde{z}$. Adjoint representation N on n is given by formula $Ad(z, t)(\zeta, \tau) = (\zeta, \tau + B(z, \zeta))$ Coadjoint: for $f \in n^*$, $g = (z, t)$, $(g \cdot f)(\zeta, \zeta) = f(\zeta, \tau) - B(z, \zeta)f(0, 1)$ then orbits for which $f|_{\tilde{z}} \neq 0$ are plane in n^* given by equation $f(0, 1) = \mu$. If $X = (\zeta, 0)$, $Y = (\zeta', 0)$, $X, Y \in n$ then symplectic structure is

$$\Omega(\sigma(\mathcal{O}, X)_f, \sigma(\mathcal{O}, Y)_f) = < f, [X, Y] > = f(0, B(\zeta, \zeta'))\mu B(\zeta, \zeta')$$

Also we have for orbit $\mathcal{O}_\mu = N/Z$ and $\mathcal{O}_\mu$ is Hamiltonian G-space.

Kirillov Character Formula or Analogy of Gabor Wavelets. Let U denote irreducible unitary representation of N with condition $U(0, t) = \exp(it\ell) \cdot 1$, where $\ell \neq 0$, then U is equivalent to representation T_ℓ which acts in $L^2(R)$ according to

$$T_\ell(z, t)\phi(x) = \exp\left(i\ell(t + px)\right)\phi(x - q)$$

If instead of N we consider E(2)/R we have S^1 case and we have Gabor functions on S^1.

Oscillator Group. Let O be an oscillator group, i.e. semidirect product of R and Heisenberg group N. Let H,P,Q,I be standard basis in Lie algebra o of the group O and H^*, P^*, Q^*, I^* be dual basis in o^*. Let functional f=(a,b,c,d) be $aI^* + bP^* + cQ^* + dH^*$. Let us consider complex polarizations $h = (H, I, P +$

iQ), $\bar{h} = (I, H, P - iQ)$ Induced from h representation, corresponding to functional f (for $a > 0$), unitary equivalent to the representation

$$W(t,n)f(y) = \exp(it(h - 1/2)) \cdot U_a(n)V(t),$$

where $\quad V(t) = \exp[-it(P^2 + Q^2)/2a], \quad P = -d/dx, \quad Q = iax,$

and $U_a(n)$ is irreducible representation of N, which have the form $U_a(z) = exp(iaz)$ on the center of N. Here we have: U(n=(x,y,z)) is Schrödinger representation, $U_t(n) = U(t(n))$ is the representation obtained from previous by automorphism (time translation) $n \longrightarrow t(n)$; $U_t(n) = U(t(n))$ is also unitary irreducible representation of N. $V(t) = \exp(it(P^2 + Q^2 + h - 1/2))$ is an operator, which according to Stone–von Neumann theorem has the property $U_t(n) = V(t)U(n)V(t)^{-1}$.

This is our last private case, but according to our approach we can construct by using methods of geometric quantization theory many "symplectic wavelet constructions" with corresponding symplectic or Poisson structure on it. Very useful particular spline–wavelet basis with uniform exponential control on stratified and nilpotent Lie groups was considered in [15].

III MELNIKOV FUNCTIONS APPROACH

In this part we continue the application of the methods of wavelet analysis to polynomial approximations of nolinear accelerator physics problems. Now we consider one problem of nontrivial dynamics related with complicated differential geometrical and topological structures of system (4). We give some points of applications of wavelet methods from the preceding parts to Melnikov approach in the theory of homoclinic chaos in perturbed Hamiltonian systems.

Routes to Chaos Now we give some points of our program of understanding routes to chaos in some Hamiltonian systems in the wavelet approach [3]-[11]. All points are:

1. A model.

2. A computer zoo. The understanding of the computer zoo.

3. A naive Melnikov function approach.

4. A naive wavelet description of (hetero) homoclinic orbits (separatrix) and quasiperiodic oscillations.

5. Symplectic Melnikov function approach.

6. Splitting of separatrix... $\longrightarrow$stochastic web with magic symmetry, Arnold diffusion and all that.

1. As a model we have two frequencies perturbations of particular case of system (4):

$$\dot{x}_1 = x_2 \quad \dot{x}_3 = x_4, \quad \dot{x}_5 = 1, \quad \dot{x}_6 = 1,$$
$$\dot{x}_2 = -ax_1 - b[\cos(rx_5) + \cos(sx_6)]x_1 - dx_1^3 - mdx_1x_3^2 - px_2 - \varphi(x_5)$$
$$\dot{x}_4 = ex_3 - f[\cos(rx_5) + \cos(sx_6)]x_3 - gx_3^3 - kx_1^2x_3 - gx_4 - \psi(x_5)$$

or in Hamiltonian form

$$\dot{x} = J \cdot \nabla H(x) + \varepsilon g(x, \Theta), \quad \dot{\Theta} = \omega, \quad (x, \Theta) \in R^4 \times T^2, \quad T^2 = S^1 \times S^1,$$

for $\varepsilon = 0$ we have:

$$\dot{x} = J \cdot \nabla H(x), \quad \dot{\Theta} = \omega \qquad (11)$$

2. For pictures and details one can see [5], [10]. The key point is the splitting of separatrix (homoclinic orbit) and transition to fractal sets on the Poincare sections.

3. For $\varepsilon = 0$ we have homoclinic orbit $\bar{x}_0(t)$ to the hyperbolic fixed point x_0. For $\varepsilon \neq 0$ we have normally hyperbolic invariant torus T_ε and condition on transversally intersection of stable and unstable manifolds $W^s(T_\varepsilon)$ and $W^u(T_\varepsilon)$ in terms of Melnikov functions $M(\Theta)$ for $\bar{x}_0(t)$.

$$M(\Theta) = \int_{-\infty}^{\infty} \nabla H(\bar{x}_0(t)) \wedge g(\bar{x}_0(t), \omega t + \Theta) dt$$

This condition has the next form:

$$M(\Theta_0) = 0, \quad \sum_{j=1}^{2} \omega_j \frac{\partial}{\partial \Theta_j} M(\Theta_0) \neq 0$$

According to the approach of Birkhoff-Smale-Wiggins we determined the region in parameter space in which we observe the chaotic behaviour [5], [10].

4. If we cannot solve equations (11) explicitly in time, then we use the wavelet approach from part 1 for the computations of homoclinic (heteroclinic) loops as the wavelet solutions of system (11). For computations of quasiperiodic Melnikov functions

$$M^{m/n}(t_0) = \int_0^{mT} DH(x_\alpha(t)) \wedge g(x_\alpha(t), t + t_0) dt$$

we used periodization of wavelet solution from part 1.

5. We also used symplectic Melnikov function approach

$$M_i(z) = \lim_{j \to \infty} \int_{-T_j^*}^{T_j} \{h_i, \hat{h}\}_{\Psi(t,z)} dt$$

$$d_i(z,\varepsilon) = h_i(z_\varepsilon^u) - h_i(z_\varepsilon^s) = \varepsilon M_i(z) + O(\varepsilon^2)$$

where $\{,\}$ is the Poisson bracket, $d_i(z,\varepsilon)$ is the Melnikov distance. So, we need symplectic invariant wavelet expressions for Poisson brackets. The computations are produced according to part 2.

6. Some hypothesis about strange symmetry of stochastic web in multi-degree-of freedom Hamiltonian systems [11].

IV SYMPLECTIC TOPOLOGY AND WAVELETS

Now we consider another type of wavelet approach which gives us a possibility to parametrize Arnold–Weinstein curves or closed loops in Hamiltonian systems by generalized refinement equations or Quadratic Mirror Filters equations.

Wavelet Parametrization in Floer Approach. Now we consider the generalization of our wavelet variational approach to the symplectic invariant calculation of closed loops in Hamiltonian systems [16]. We also have the parametrization of our solution by some reduced algebraical problem but in contrast to the general case where the solution is parametrized by construction based on scalar refinement equation, in symplectic case we have parametrization of the solution by matrix problems – Quadratic Mirror Filters equations [17].

The action functional for loops in the phase space is [16]

$$F(\gamma) = \int_\gamma pdq - \int_0^1 H(t, \gamma(t))dt$$

The critical points of F are those loops γ, which solve the Hamiltonian equations associated with the Hamiltonian H and hence are periodic orbits. By the way, all critical points of F are the saddle points of infinite Morse index, but surprisingly this approach is very effective. This will be demonstrated using several variational techniques starting from minimax due to Rabinowitz and ending with Floer homology. So, (M,ω) is symplectic manifolds, $H : M \to R$, H is Hamiltonian, X_H is unique Hamiltonian vector field defined by

$$\omega(X_H(x), v) = -dH(x)(v), \quad v \in T_x M, \quad x \in M,$$

where ω is the symplectic structure. A T-periodic solution $x(t)$ of the Hamiltonian equations

$$\dot{x} = X_H(x) \quad \text{on } M$$

is a solution, satisfying the boundary conditions $x(T) = x(0), T > 0$. Let us consider the loop space $\Omega = C^\infty(S^1, R^{2n})$, where $S^1 = R/\mathbf{Z}$, of smooth loops in R^{2n}. Let us define a function $\Phi : \Omega \to R$ by setting

$$\Phi(x) = \int_0^1 \frac{1}{2} < -J\dot{x}, x > dt - \int_0^1 H(x(t))dt, \quad x \in \Omega$$

The critical points of Φ are the periodic solutions of $\dot{x} = X_H(x)$. Computing the derivative at $x \in \Omega$ in the direction of $y \in \Omega$, we find

$$\Phi'(x)(y) = \frac{d}{d\epsilon}\Phi(x + \epsilon y)|_{\epsilon=0} = \int_0^1 < -J\dot{x} - \nabla H(x), y > dt$$

Consequently, $\Phi'(x)(y) = 0$ for all $y \in \Omega$ iff the loop x satisfies the equation

$$-J\dot{x}(t) - \nabla H(x(t)) = 0,$$

i.e. $x(t)$ is a solution of the Hamiltonian equations, which also satisfies $x(0) = x(1)$, i.e. periodic of period 1. Periodic loops may be represented by their Fourier series:

$$x(t) = \sum_{k \in \mathbf{Z}} e^{k2\pi Jt} x_k, \quad x_k \in R^{2k},$$

where J is quasicomplex structure. We give relations between quasicomplex structure and wavelets in [11]. But now we use the construction [17] for loop parametrization. It is based on the theorem about explicit bijection between the Quadratic Mirror Filters (QMF) and the whole loop group: $LG : S^1 \to G$. In particular case we have relation between **QMF**-systems and measurable functions $\chi : S^1 \to U(2)$ satisfying

$$\chi(\omega + \pi) = \chi(\omega)\begin{bmatrix} 0 & 1 \\ 1 & 0 \end{bmatrix},$$

in the next explicit form

$$\begin{bmatrix} \hat{\Phi}_0(\omega) & \hat{\Phi}_0(\omega + \pi) \\ \hat{\Phi}_1(\omega) & \hat{\Phi}_1(\omega + \pi) \end{bmatrix} = \chi(\omega)\begin{bmatrix} 0 & 1 \\ 1 & 0 \end{bmatrix} + \chi(\omega + \pi)\begin{bmatrix} 0 & 0 \\ 0 & 1 \end{bmatrix},$$

where

$$\left|\hat{\Phi}_i(\omega)\right|^2 + \left|\hat{\Phi}_i(\omega + \pi)\right|^2 = 2, \quad i = 0, 1.$$

Also, we have symplectic structure on LG

$$\omega(\xi, \eta) = \frac{1}{2\pi}\int_0^{2\pi} < \xi(\theta), \eta'(\theta) > d\theta$$

So, we have the parametrization of periodic orbits (Arnold–Weinstein curves) by reduced QMF equations.

Extended version and related results may be found in [3]-[11].

One of us (M.G.Z.) would like to thank A. Dragt, J. Irwin, F. Schmidt for discussions, Zohreh Parsa for many discussions and continued encouragement during and after workshop "New Ideas for Particle Accelerators" and Institute for Theoretical Physics, University of California, Santa Barbara for hospitality.

This research was supported in part under "New Ideas for Particle Accelerators Program" NSF- Grant No. PHY94-07194.

REFERENCES

1. Dragt, A.J., *Lectures on Nonlinear Dynamics*: CTP, 1996.
2. Heinemann, K., Ripken, G., Schmidt, F.: DESY 95-189, 1995
   Ripken G., Schmidt F.: CERN/SL/95-12(AP) DESY 95-063, 1995.
3. Fedorova A.N., Zeitlin M.G.: Proc. of 22 Summer School'Nonlinear Oscillations in Mechanical Systems' St. Petersburg, 1995, p. 89.
4. Fedorova, A.N., and Zeitlin, M.G.: Proc. of 22 Summer School'Nonlinear Oscillations in Mechanical Systems' St. Petersburg, 1995, p. 97.
5. Fedorova, A.N., and Zeitlin, M.G.: Proc. of 22 Summer School'Nonlinear Oscillations in Mechanical Systems' St. Petersburg, 1995, p. 107.
6. Fedorova, A.N., and Zeitlin, M.G.: Proc. of 23 Summer School 'Nonlinear Oscillations in Mechanical Systems' St. Petersburg, 1996, p. 322.
7. Fedorova, A.N., and Zeitlin, M.G.: Proc. 7th IEEE DSP Workshop, Norway, 1996, p. 409.
8. Fedorova, A.N., and Zeitlin, M.G.: Proc. 2nd IMACS Symp. on Math. Modelling, ARGESIM Report **11**, Austria (1997) 1083.
9. Fedorova, A.N., and Zeitlin, M.G.: EUROMECH-2nd European Nonlinear Oscillations Conf. (1997) 79.
10. Fedorova, A.N., and Zeitlin, M.G.: EUROMECH-2nd European Nonlinear Oscillations Conf. (1997) 153.
11. Fedorova, A.N., and Zeitlin, M.G.: Proc. of 24 Summer School'Nonlinear Oscillations in Mechanical Systems' St. Petersburg (1997).
12. Latto, A., Resnikoff, H.L., and Tenenbaum, E.: Aware Technical Report AD910708 (1991).
13. Kalisa, C., and Torresani, B., *N-dimensional Affine Weyl-Heisenberg Wavelets*: preprint CPT-92 P.2811 Marseille, 1992.
14. Kawazoe, T.: Proc. Japan Acad. **71** Ser. A, 1995, p. 154.
15. Lemarie P.G.: Proc. Int. Math. Congr., Satellite Symp., 1991, p. 154.
16. Hofer E., Zehnder E., *Symplectic Topology*: Birkhauser, 1994.
17. Holschneider M., Pinkall U.: CPT-94 P.3017, Marseille, 1994.

Some Problems of Nonlinear Aberration Correction

Serge N. Andrianov

St. Petersburg State University, St. Petersburg, 198904, Russia
E-mail:serge@asn.apmath.spb.su

Abstract

Abstract. This paper is devoted to one of important problems of beam physics: correction of nonlinear harmful aberrations in different kind of beam lines. There are a set of approaches to solving this problem. In this paper we consider an approach based on algebraic description of nonlinear aberrations. In particular, this approach allows to reduce the problem of correcting fields searching to linear algebraic problems. For this purpose the matrix formalism for Lie algebraic tools is used. This permits to create databases of symbolic matrices which can be used for the correcting procedure. [1]

INTRODUCTION

In this report we consider more the mathematical and computational problems than physical one. Indeed, from the physical point of view the problem of nonlinear aberrations correction has clear meaning: we must maximally decrease all harmful aberrations, including geometrical and chromatic aberrations, influence of fringing fields, aberrations induced by space charge forces. But in any particular case one deals with concrete machine with its own characteristics and properties. That is why a creation of calculation algorithms for this problem must take account to these particular points. It is better if the used mathematical methods allow this procedure.

LIE TRANSFORMATIONS

It is known that time evolution of dynamical systems may be represented by one-parameter group of maps acting on the initial values of phase space variables $\mathcal{M} : X_0 \to X = \mathcal{M} \circ X_0$. In the case of Hamiltonian systems it is

---

[1] This work is supported by the Russian Foundation for Fundamental Researches 96-02-17335-a

a symplectic group of canonical maps – Lie maps. For general case similar maps will be called Lie transformations.

The Basic definitions and concepts

First of all, we give a brief essay of the matrix formalism for Lie algebraic tools and its computer realization [1]-[5].

Lie Transformations generated by Ordinary Differential Equations

Let $F(X, U; t)$ be an analytical function on phase vector X, where $X \in \mathfrak{X}$, $\dim X = n$, $s \in [s_0, S]$, $U \in \mathfrak{U}$ – a vector of control parameters, $\mathfrak{X}$, $\mathfrak{U}$ are corresponding spaces and the following ordinary differential equation is a motion equation for particles in a beam line (s – is an independent variable measured along a reference orbit)

$$\frac{dX}{ds} = F(X, U; s). \qquad (1)$$

In general, any solution of this equation can be written in the form (here problems of existence and convergence here are not discussed)

$$X(X_0, U; s|s_0) = \mathcal{M}(U; s|s_0) \circ X_0, \quad X_0 = X(s_0), \qquad (2)$$

where $\mathcal{M}$ is a Lie transformation generated by the right side of the Eq.(1). The Eq.(2) gives a nonlinear mapping from an initial state X_0 to a final (or current) state $X = X(s)$. It is known that this map can be represented as a chronological exponent operator (or Dyson operator)

$$\mathcal{M}(F, U; s|s_0) = \mathrm{T} \exp\left\{ \int_{s_0}^{s} \mathcal{L}_{F(X,U;\tau)} d\tau \right\}.$$

Using the Magnus representation [6] we can write

$$\mathcal{M}(F, U; s|s_0) = \exp\left\{ \mathcal{L}_{G(X,U;s|s_0)} \right\}. \qquad (3)$$

Here $\mathcal{L}_F$, $\mathcal{L}_G$ are Lie operators associated with vector functions F and G. The function G can be calculated with the help of the continuous analog of the CBH formula.

For the function $F(X, U; s)$ we can write an expansion

$$F(X, U; s) = \sum_{k=0}^{\infty} F_k(X, U; s), \quad (4)$$

where $F_k(X, U; s)$ are homogeneous polynomials of X. This representation generates a similar representation for the function G in the Magnus representation:

$$G(X, U; s|s_0) = \sum_{k=0}^{\infty} G_k(X, U; s|s_0).$$

Using the Eq.(4) one can write for the Lie operator $\mathcal{L}_F$

$$\mathcal{L}_{F(X,U;s)} = \sum_{k=0}^{\infty} F_k(X, U; s)^* \frac{\partial}{\partial X} = \sum_{k=0}^{\infty} \mathcal{L}_{F_k(X,U;s)}. \quad (5)$$

The Lie transformation $\mathcal{M}$ (for example, in the form of the Eq.(3)) can be represented in a factorized form [7]–[8]

$$\mathcal{M}(F; s|s_0) = \ldots \circ \mathcal{M}(\hat{G}_k) \circ \ldots \circ \mathcal{M}(\hat{G}_2) \circ \mathcal{M}(\hat{G}_1) =$$

$$\mathcal{M}(\tilde{G}_1) \circ \mathcal{M}(\tilde{G}_2 \circ \ldots \circ \mathcal{M}(\tilde{G}_k) \circ \ldots. \quad (6)$$

In autonomous case $(\partial F/\partial s = 0)$ we can present the functions $\hat{G}_k$, where $G = \sum_{k=0}^{\infty} G_k$, as sums of two terms: $G_k = G_k^k + G_k^{\{<k\}}$, where $G_k = G(\tau, F_k)$, $G_k^{\{<k\}} = G_k(\tau, \{F_j\}; j < k)$, $\tau = s - s_0$. One can obtain the following representation for G_k^k

$$G_k^k = \sum_{j=0}^{\infty} \frac{\tau^k}{j!} \mathcal{L}_{G_1}^{(j-1)} \circ F_k,$$

For the second term there is not such simple form. But, using the Zassenhauss formula we can obtain necessary relations for any order of τ. Note that $G_2^{\{<2\}} = 0$. As an example one can obtain for $G_3^{\{<3\}}$

$$G_3^{\{<3\}} = -\frac{\tau^3}{4} \left[\frac{1}{3} \mathcal{L}_{F_2} + \frac{\tau}{6} \mathcal{L}_{F_2} \circ \mathcal{L}_{F_1} + \frac{\tau^2}{20} \mathcal{L}_{F_2} \circ \mathcal{L}_{F_1}^2 + \frac{\tau^2}{30} \mathcal{L}_{\tilde{F}_2} \circ \mathcal{L}_{F_1} \right] \circ \tilde{F}_2,$$

$$\tilde{F}_2 = \mathcal{L}_{F_1} \circ F_2.$$

The Matrix Representation: Basic Definitions

For the homogeneous polynomials $F_k(X,U;s)$ we can write a representation

$$F_k(X,U;s) = \mathbf{F}_k(U;s)X^{[k]}, \qquad (7)$$

where $\mathbf{F}_k(U;s)$ are matrices $\left(n \times \binom{n+k-1}{k}\right)$. For the function G_k (for the Magnus representation):

$$G_k(X,U;s|s_0) = \mathbf{G}_k(U;s|s_0)X^{[k]}. \qquad (8)$$

Here $X^{[k]} = \underbrace{X \otimes \ldots \otimes X}_{k-times}$ is the Kronecker power for X of k-th order.
Corresponding matrix representations for G_k^k and $G_k^{\{<k\}}$ can br written with the help of corresponding matrices

$$\mathbf{G}_k^k = \mathbf{G}_k^k(\tau, \mathbf{F}_k) = \sum_{j=1}^{\infty} \frac{\tau^j}{j!} \mathbf{F}_k \left(\mathbf{G}_1^{\oplus k}\right)^{(j-1)}.$$

For our example for $\mathbf{G}_3^{\{<3\}}$ one can write

$$\mathbf{G}_3^{\{<3\}} = -\frac{\tau^3}{4}\tilde{\mathbf{F}}_2 \left[\frac{1}{3}\mathbf{F}_2^{\oplus 2} + \frac{\tau}{6}\hat{\mathbf{F}}_2 + \frac{\tau^2}{20}\mathbf{F}_1^{\oplus 2}\hat{\mathbf{F}}_2 + \frac{\tau^2}{30}\mathbf{F}_1^{\oplus 2}\left(\tilde{\mathbf{F}}_2\right)^{\oplus 2} \right],$$

$$\tilde{\mathbf{F}}_2 = \mathbf{F}_2\mathbf{F}_1^{\oplus 2}, \qquad \hat{\mathbf{F}}_2 = \mathbf{F}_1^{\oplus 2}\mathbf{F}_2^{\oplus 2},$$

where $\mathbf{A}^{\oplus k}$ denotes the k-multiple Kronecker sum for a matrix $\mathbf{A}$.

The representations (6)–(8) allow step–by–step to operate by Lie transformation associated with any homogeneous polynomial of m-th order on the phase vector X. For this purpose it is convenient to use the following expression

$$\mathcal{M}(G_m) \circ X^{[l]} = \exp\{\mathcal{L}_{G_m}\} \circ X^{[l]} = X^{[l]} + \sum_{k=1}^{\infty} \frac{\mathbf{P}_m^{kl}}{k!} X^{[k(m-1)+l]},$$

$$\mathbf{P}_m^{kl} = \prod_{j=1}^{k} \mathbf{G}_m^{\oplus\,(j-1)(m-1)+l}.$$

Define $\mathcal{M}_{\leq k}$ as a finite product

$$\mathcal{M}_{\leq k} = \mathcal{M}(G_k) \circ \ldots \circ \mathcal{M}(G_2) \circ \mathcal{M}(G_1),$$

than, for example,

$$\mathcal{M}_{\leq 3} \circ X = \mathbf{M}^{11}\left(X + \sum_{m=2}^{3}\sum_{k=1}^{\infty} \frac{\mathbf{P}_m^{k1}}{m!} X^{[k(m-1)+1]} + \right.$$

$$\left.\sum_{l=1}^{\infty}\sum_{k=1}^{\infty} \frac{1}{k!m!} \mathbf{P}_2^{k1} \mathbf{P}_3^{l(k+1)} X^{[2l+k+1]}\right) \approx$$

$$\mathbf{M}^{11}\left(X + \mathbf{P}_2^{11} X^{[2]} + \left(\mathbf{P}_3^{11} + \frac{1}{2!}\mathbf{P}_2^{21}\right) X^{[3]}\right).$$

We can build the following matrices

$$\mathbf{M}^{ml} = \sum_{\substack{k_1+\ldots+k_m=l \\ k_i \geq 1}} \bigotimes_{i=1}^{m} \mathbf{M}^{1 k_i}, \quad l \geq m.$$

All these matrices can be combined into an infinite dimensional matrix $\mathbf{M}^\infty$:

$$\mathbf{M}^\infty = \begin{pmatrix} \mathbf{M}^{11} & \mathbf{M}^{12} & \ldots & \mathbf{M}^{1k} & \ldots \\ 0 & \mathbf{M}^{22} & \ldots & \mathbf{M}^{2k} & \ldots \\ \vdots & \vdots & \ddots & \vdots & \ddots \\ 0 & 0 & \ldots & \mathbf{M}^{kk} & \ldots \\ \vdots & \vdots & \ddots & \vdots & \ddots \end{pmatrix}.$$

And there is an algebraic equation

$$X^\infty = \mathbf{M}^\infty \cdot X_0^\infty,$$

where $X^\infty = \left(X\ X^{[2]} \ldots X^{[k]} \ldots\right)^*$. For the inverse matrix $\mathbf{T}^\infty = (\mathbf{M}^\infty)^{-1}$ we have the similar uptriangular structure and its block–matrices can be calculated according the following relations (the generalized Gauss's algorithm)

$$\mathbf{T}^{kk} = \left(\mathbf{M}^{kk}\right)^{-1} = \left(\mathbf{M}^{11}\right)^{-[k]},$$

$$\mathbf{T}^{ik} = -\mathbf{T}^{ii} \sum_{j=i+1}^{k} \mathbf{M}^{ij}\mathbf{T}^{jk}, \quad i<k, \quad \mathbf{T}^{ik} \equiv \mathbf{0}, \quad i>k.$$

All calculations are carried out with the help of computer algebra codes (in this work *REDUCE* and *MAPLE* codes are used). For manipulating by these relations some databases are created (they consist of block–matrices $\mathbf{M}^{ik}$ and $\mathbf{T}^{ik}$). The auxiliary databases contain ready formulae for noncommutative variables (for example, the CBH formula up to some orders) (see, [7],[9]–[11]).

NONLINEAR ABERRATIONS

An Investigation Step

So, we put the matrix formalism into nonlinear aberrations study. According to the Lie algebra approach each distinct aberration corresponds to an unique monomial in the Lie operator (5) which is the generator for a beam line. This leads to close connection between an aberrations set and their mapping in terms of elements of the matrices $\mathbf{F}_k$, $k \geq 1$. Each matrix $\mathbf{F}_k$ collects all aberrations of an identical order (k-th order). There are two ways. The first is based on a representation of set of aberrations as a set of "elementary aberrations" and the second way is based on a set of physical aberrations. We should note that aberrations of the first type usually have an abstract description without connection with physical sources of aberrations. They have very simple structures. This type of aberrations is more general and can be used for description of different kind of dynamical systems. Moreover, in this case the corresponding matrices $\mathbf{F}_k \in \{\mathbf{F}_k\}$ also have simple structures and the matrices $\mathbf{M}_k\left(\{\mathbf{F}_j\}_{j \leq k}\right)$ can be calculated in symbolic form and protected in one of databases. Any physical aberration can be designed from these "elementary aberrations" as from bricks. For this purpose one can use oriented graphs methods on two steps: for creating designed physical aberrations from "elementary aberrations" and than for finding corresponding solving matrix $\mathbf{M}^{1k}$.

On the contrary, aberrations of the second type have simple physical interpretations (for example, geometrical, chromatic and so on). This picture of description is very close to well known classification of an aberration set in light optics (see, for example, [12]). For beam dynamics problems such classification is more complete and have not be done yet. But there exists a set of important aberrations which play the main role for beam line design. For these aberrations we also have a set of known matrices $\{\mathbf{F}_k\}$ and corresponding set of solving matrices $\mathbf{M}^{1k}\left(\{\mathbf{F}_j\}, j \leq k\right)$. The simplest form of such kind of representataion we have for the kick approximation. In this case both ways are very close to each other. For an investigation of influence of aberrations for particular beam line is very convenient to use a set of weight coefficients for each element (or for a group of elements) of corresponding matrices. This allows to investigate a virtual "influence" of every element on desired beam characteristics. As an example, it may be mentioned the problem of optimal ion–optical system (for example, microprobe systems [13]). For such system so called spherical aberrations (one type of geometrical aberrations) play a main role. Both analytical study and above mentioned approach demonstrate the influence only four elements of the matrix $\mathbf{M}^{13}$, which are coefficients for following third order monomials: x'^3, $x'y'^2$, y'^3 and $y'x'^2$. Moreover, the pointed approach allows to estimate necessary limits for the elements $\mathbf{M}^{13}$. The similar approach was applied for an invisible turning insertions modelling. For this problem we studied chromatic aberrations and space charge induced aberrations too. Besides, there are some beam physics problem connected with a presence of some kind of harmful aberrations. As an example the induced (parasitic) oscillations of an intensity of extracting from circular accelerators beam can be discussed [14]. At first, for this problem one have to study an influence of different kind of source of such oscillations. Than it is necessary to include a feedback (which can to minimize the oscillations) to the calculation procedure and at last to determine necessary parameters of such feedback. For slow extraction system (in JINR, Dubna, Russia) such investigation was carried out (see references in [14]).

Here we should mention another approaches to study an influence of different kind of nonlinear aberrations on beam characteristics. This approach is based on concepts of nonlinear invariants for a control beam system. In the paper [4] some methods of nonlinear invariants construction are discussed. A concept of kinematic invariants [4] for linear dynamical systems is considered as a base of this approach. But for more effective usage of this approach one must computing the beam line carefully.

Algebraic Methods for Aberrations Correction

As an example, let demonstrate one of approaches to define corrector (octupole forces) values on the example for the microprobe system [15]. According to our approach we can write up to third order

$$X = \mathbf{M}^{11} X_0 + \mathbf{M}^{13} X_0^{[3]},$$

where the matrix $\mathbf{M}^{13}$ describes the nonlinear aberrations of third order generated by quadrupoles and octupoles. For $\mathbf{M}^{13}$ one can write

$$\mathbf{M}^{13}(S|s_0) = \mathbf{M}^{11}(S|s_0)\mathbf{Q}^{13}(S|s_0) =$$

$$\mathbf{M}^{11}(S|s_0) \left(\int_{s_0}^{S} \left(\mathbf{M}^{11}(\tau|s_0)\right)^{-1} \mathbf{P}^{13}(\tau) \mathbf{M}^{33}(\tau|s_0) \right) =$$

$$\mathbf{M}^{11}(S|s_0) \sum_{k=1}^{NO} \mathbf{Q}^{13}(s_{k-1}|s_k),$$

where NO is a number of compensating octupoles on the total length of the microprobe $(S - s_0)$, $\mathbf{P}^{13}(s)$ is a matrix describing of the control field (for quadrupoles and octupoles) distribution along the beam axis. We can write

$$\mathbf{P}^{13} = \mathbf{P}^{13}_{quadr} + \mathbf{P}^{13}_{oct}(\Gamma),$$

where $\Gamma = (\gamma_1, \ldots, \gamma_{NO})^*$ is a vector of octupole forces γ_i. Than the matrix $\mathbf{Q}^{13}$ can be rewritten as

$$\mathbf{Q}^{13}(S|s_0) = \int_{s_0}^{S} \left(\mathbf{M}^{11}(\tau|s_0)\right)^{-1} \mathbf{P}^{13}_{quadr}(\tau)\mathbf{M}^{33}(\tau|s_0) +$$

$$\sum_{k=1}^{NO} \int_{s_{k-1}}^{s_k} \left(\mathbf{M}^{11}(\tau|s_0)\right)^{-1} \mathbf{P}^{13}_{k-oct}(\tau)\mathbf{M}^{33}(\tau|s_0) =$$

$$\mathbf{Q}^{13}_{quadr}(S|s_0) + \sum_{k=1}^{NO} \gamma_k \int_{s_{k-1}}^{s_k} \mathbf{M}^{11}(s_0|\tau)\mathbf{P}^{13}_{k-oct}(\tau)\mathbf{M}^{33}(\tau|s_0),$$

where $\mathbf{P}_{k-oct}$ is a matrix calculated under the condition $\gamma_k = 1$, $\gamma_j = 0$, $\forall j \neq k$. This is identical for all intervals $[s_{k-1}, s_k]$ (it can be part of a corresponding database). Define

$$\mathbf{Q}^{13}_{k-oct}(s_k|s_{k-1}) = \gamma_k \int_{s_{k-1}}^{s_k} \mathbf{M}^{11}(s_0|\tau)\mathbf{P}^{13}_{k-oct}(\tau)\mathbf{M}^{33}(\tau|s_0).$$

Finally, one can write

$$X = \mathbf{M}^{11}\left(X_0 + \left[\mathbf{Q}^{13}_{quadr} + \sum_{k=1}^{NO} \mathbf{Q}^{13}_{k-oct}\right] X_0^{[3]}\right).$$

For the JINR microprobe system NO = 4 and we can write the following linear algebraic equation for defining of γ_k, $k = \overline{1,4}$:

$$\mathbf{A}\Gamma = B_{quadr}, \qquad (9)$$

where the matrix $\mathbf{A}$ consist of the $\mathbf{M}^{13} = \mathbf{M}^{11}\mathbf{Q}^{13}(S|s_0)$ elements:

$$\mathbf{A} = \begin{pmatrix} r_{14}(1000) & r_{14}(0100) & r_{14}(0010) & r_{14}(0001) \\ r_{1\,10}(1000) & r_{1\,10}(0100) & r_{1\,10}(0010) & r_{1\,10}(0001) \\ r_{44}(1000) & r_{44}(0100) & r_{44}(0010) & r_{44}(0001) \\ r_{4\,10}(1000) & r_{4\,10}(0100) & r_{4\,10}(0010) & r_{4\,10}(0001) \end{pmatrix},$$

and
$$B_{quadr} = (r_{14}(0000)\, r_{1\,10}(0000)\, r_{44}(0000)\, r_{4\,10}(0000))^*.$$

Here r_{ik} are elements of the $\mathbf{M}^{13}$. From the Eq. (7) one can find compensating octupole values $\{\gamma_k\}_{k=\overline{1,4}}$. Note that the solving process of the Eq.(6) can be realized in nonlinear programming codes. This approach allows to define a sensitivity of beam characteristics while a calculation process. Moreover, a calculation accuracy must be in an agreement with a tolerances precision. This give a possibility to operate the design process and to give corresponding recommendations. For the JINR microprobe system nonlinear aberrations lead to increasing of sizes of the beam image on a target from 1μ for the linear model up to 6μ for third order aberrations. After compensating procedure we obtained 1μ again [16]. Note that similar procedure were carried out for fifth order too. Taking into account this order led to not large growth of the terminal envelope values (up to 1.3μ). That is why the procedure of aberrations compensating were not necessary. But for another cases one can operate by similar way.

Optimal Control Theory and Nonlinear Aberrations

There is another approach to solve the aberrations correction problems. This approach is based on methods of the optimal control theory. For this one can choose an admissible set of control functions (in our case they are multipole force distributions along the system optical axis) and construct a necessary object function and a set of boundary conditions. The using one of methods of optimal control theory we can define optimal distribution functions ensuring the desired compensating fields. In this work for this purpose the control theory problem is reduced to a nonlinear programming problem. Parameters determining required functions are replaced with starting from physical meaning of corresponding functions. For example, for quadrupole gradient distribution function $g(s)$ one can select one of a suitable model function depending on some set of parameters (an effective length, a maximal value of the gradient, parameters of fringing fields and so on). The residual function between desired size of the beam image and a current size is the objective function for optimization. This approach is very useful when one wants to use a few correcting multipole lenses, because the corresponding linear equation (see the Eq.(7)) will be underdetermined in this case.

Feedback Construction

For some beam physics problems the control criterion function $F[U]$ has a time structure (here U is an admissible control vector):

$$F[U](t|t_0) = \int_{t_0}^{t} \int_{\mathfrak{M}_\tau} G(X(\tau), U(\tau), \tau) dX d\tau, \quad (10)$$

where G is appropriate function describing an investigated physical phenomenon, $\mathfrak{M}_\tau$ is a phase set occupied by particles at some current moment τ. For example, the time structure of a beam extracted from circular accelerator. The source of harmful aberrations of the extracted beam is unstubility of feed currents in different magnet windings. This unstability is unremovable and that is why we should introduce a feedback action for compensation of an influence of the current instability. It is obviously that the beam evolution process has an inertia measure which has to be included into our investigation. For demonstration of our approach we can confine ourselves to second order approximation. In this case one can write

$$X = \mathbf{M}^{11}(t|t_0) + \left(\mathbf{M}^{12}_{com}(t|t_0) + \mathbf{M}^{12}_{fb}(t|t_0)\right) X_0^{[2]}, \quad (11)$$

where $\mathbf{M}^{12}_{com}(t|t_0)$ is the matrix describing an action of the general fields and $\mathbf{M}^{12}_{fb}(t|t_0)$ is the matrix for feedback description. It is necessary note that the length of the feedback impulse τ_{fb} plays a main role in the compensating process. This role is defined by the inertia of the beam. For modelling of compensating process we should consider a beam evolution for some hundred turns only for one feedback impulse. That is why the problem of effectiveness of used mathematical models is very important. The adiabatic time evolution for slow extract process and the used matrix formalism for Lie transformations (3) allow to calculate necessary time depended matrix $\mathbf{M}^{11}$, $\mathbf{M}^{12}_{com}$, $\mathbf{M}^{12}_{fb}$ according to the following rule:

$$\exp\{(t - t_0)\mathbf{A}\} = \exp\{N\Delta t \mathbf{A}\} = (\exp\{\Delta t \mathbf{A}\})^N.$$

The value of Δt should be chosen from a precision condition for the calculation process. This rule allows to decrease calculation time. For the JINR synchrotron slow extract process we consider time dependence for all parameters of the corresponding motion equation (see [13]), including parasitic oscillations of magnet fields of different nature. The object function (8) for this problem describes a modulation coefficient K_{mod} for extracted beam. The optimization process permits to decrease the value of K_{mod} from

30 – 40% up to 3 – 5% for feedback impulses with realizable characteristics [14].

Organization of Calculations

It is known that almost all calculations of dynamical system behavior, in particularly, for beam physics, have to deal with numerical and symbolical methods. These calculations are tedious and time consuming. It is especially hard to be certain of long time evolution problems. That is why the choice of mathematical methods plays a great role. Present range of codes for beam line design is akin to a jungle. This variety as variety of mathematical methods for solving corresponding problems lead to practical impossibility to create compact a codes system for modelling permitting natural extension as new knowledge accumulation. For this aim we use an approach based on object–oriented concepts. All levels of a modelling process use dynamic modeling paradigm [2]. This means that a researcher can manipulate by mathematical methods and by physical models as objects. For this goal we provide the necessary software with an intellectual interface and describe calculation experiments in the terms of expert systems modules. As one of the basic using mathematical methods is the matrix formalism for Lie algebraic transformations than usage of this method allows to create databases and knowledge bases basic contents of which are matrices in symbolic forms [11]. The corresponding calculations are made with the help of computer algebra codes. From beam line designer's viewpoint these matrices are a collection of *bricks* from which he can design his "building". These *bricks* have regular nature as corresponding operations are also regular than the designer's tools should permit to include or extract a model of a beam line. The structure of solving module of the expert system is based on a synthesis of knowledge of different models for beam lines and solving methods (see a scheme on the Fig.1). On this scheme we present basic submodules for the solving module, which is intended for analysis and synthesis procedures for the beam line under study. The uniformity of databases elements and operations for their manipulating helps to create this solving module. The natural language interface uses a user's menu, which can be created with the help such visual programming codes as *DELPHI*. The structure of the interface must give maximal comfortable (with necessary help) tools for designing the beam line. For this purpose we foresee a collection of stirring elements (in the form of a database) which can be enlarged as required. The computer algebra (CA) module is intended for the supplement procedures.

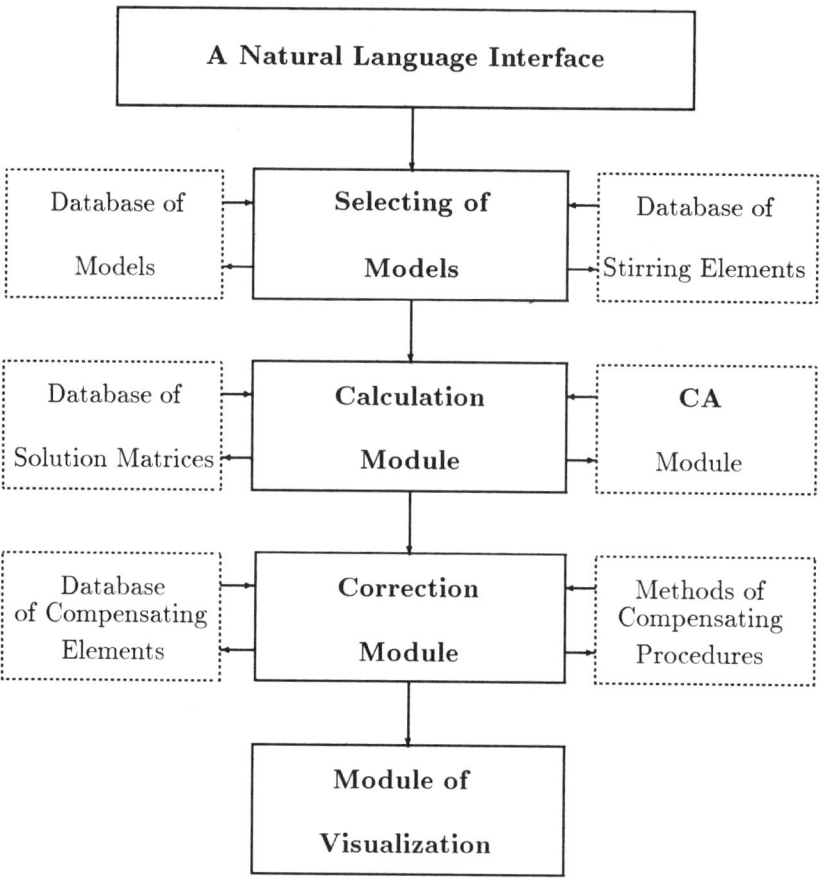

Fig.1

This approach allows to applicate one or more methods for careful investigation of nonlinear dynamics, to compare results and to choose the best method. In particular, a researcher can choose solving methods, modify them (if it necessary) [17], creates invariants [4] and so on. All operations can be made in symbolic forms with the help special computer algebra packages (realized in *REDUCE* or *MAPLE* codes). A necessary methods selection one can from *Method Databases*. Then one runs calculation process. From the point of the matrix formalism one uses a set of ready formulae in symbolic form for matrices elements in dependence on his problem. The necessary solution information are contained as a set of solution matrices $\mathbf{M}^{1k}$, which can be calculated in advance in symbolic forms and are kept in corresponding databases. For extraction process the oriented graphs methods are used. Manipulations by these matrices allows to realized the dynamic modeling paradigm. All these calculation steps are surrounded by corresponding interface which allows to realize the modelling process in natural (for given object problem) terms. In particular, for this purpose the *DELPHI* codes can be used. This visual programming environment uses the object–oriented ideology to a considerable extent. The designed *Windows* applets can be filled up as growth of our knowledge and experience. If it necessary one can connect both other beam physics packages using *OLE* (*Object Linking embedding*) technology and general mathematical packages. Here we would like to point out a main property of this suggested approach: the opportunity to work with abstract objects. This approach allows to cross from problems of analysis of beam lines to problem of synthesis of machines with desired characteristics.

REFERENCES

1. Andrianov S.N. *A Matrix Representation of the Lie Transformation*, Proc. of the Int. Congress on Computer Systems and Applied Mathematics, CSAM'93, St.Petersburg, July 19-23, 1993, St.Petersburg, p.14.

2. Andrianov S.N. *Dynamic Modeling in Beam Dynamics*, Proc. of the Second Int. Workshop *Beam Dynamics & Optimization BDO'95*, St.Petersburg, July 4-8, 1995, St.Petersburg State University, St.Petersburg, 1995, pp. 25–32.

3. Andrianov S.N. *Analytical Simulation of Space Charge in Ion–Optical Systems*, Proc. of the First Int. Workshop *Beam Dynamics & Optimization BDO'94*, St.Petersburg, July 4-8, 1994, St.Petersburg State University, St.Petersburg, 1995, pp. 19–29.

4. Andrianov S.N. *Construction of Approximate Symmetries and Invariants for Dynamical Systems*, Proc. of the Second Int. Workshop *Beam Dynamics & Optimization BDO'95*, St.Petersburg, July 4-8, 1995, St.Petersburg State University, St.Petersburg, 1995, pp. 16–24.

5. Andrianov S.N. *The Group-Theoretical and Algebraic Modelling of Particle Beam Control Systems*, Mechanics and Control Problems, issue 15, The Theory of Control Systems, St.Petersburg, St.Petersburg State University, 1992, pp.7–13.

6. Magnus W. *On the Exponential Solution of Differential Equations for a Linear Operator*, Comm. on Pure and Appl. Math.,7,No 4, pp. 649-673, 1954.

7. Dragt A.J. *Lectures on Nonlinear Orbit Dynamics*, Physics of High Energy Particle Accelerators, ed. by R.A.Carrigan et al., AIP Conf. Proc., **87**, 1982, pp.147-313.

8. Gajaja I. *Monomial Factorization of Symplectic Maps*, Particle Accelerators, **43**, N 3, 1994, pp.133–144.

9. Koseleff P–V. *Formal Calculus for Lie Methods in Hamiltonian mechanics*, Ph.D. Thesis, Ecole Polytechnique (1993).

10. Gerdt V.P., Akselrod I.R., Kovtun V.E., Robuk V.N. *Construction of a Lie Algebra by a Subset of Generators and Commutation Relations*, Computer Algebra in Physical Researches – Singapore: World Scientific, 1991.– pp.306–312.

11. Chaklerov O.G. *Automatization of the Motion Equations Creation for Multipole*

Magnet Systems, Proc. of the Third Int. Workshop *Beam Dynamics & Optimization BDO'96*, St.Petersburg, July 1–5, 1995, St.Petersburg State University, will be published.

12. Krötzsch G., Wolf K.B. *Group-Classified Polynomials of Phase Space in Higher-Order Aberration Expansions*, Comunicaciones Técnicas, Inst. de Invest.en Matematicas Aplicadas y en Sistemas, Univ.Nac.Autonoma de Mexico, N 563, 1990, Mexico.

13. Andrianov S.N., Yudin I.P. *Nuclear Microprobe System with Defined Characteristics*, Proc. of XIII Conf. on Charge Particle Accelerators, JINR, Dubna, Oct.13-15, 1992, Dubna (Russia), part 2, pp.305-309, 1993 (in Russian).

14. Andrianov S.N. *Some Problems of Resonance Beam Extraction*, Proc. of the First Int. Workshop *Beam Dynamics & Optimization BDO'94*, St.Petersburg, July 4-8, 1994, St.Petersburg State University, St.Petersburg, 1995, pp. 31–33.

15. Dragt A.J., Gluckstern R.L., Neri F., Rangarajan G. *Theory of emittance invariants*, Frontiers of Particles Beams; Observation, Diagnosis and Correction, ed. by M.Month, S.Turner, Lectures Notes in Physics, **343**, 1989, pp.94-121.

16. Andrianov S.N., Dymnikov A.D., Osetinsky G.M. *Correction of Geometrical Aberrations in the Proton Microzond from Quadrupole Lenses*, JINR, Dubna, 1984, B-1-9-84-209, p.8.

17. Andrianov S.N.,Dvoeglazov A.I. *A Solving Module for Hamiltonian Dynamical Systems*, Proc. of the Third Int. Workshop *Beam Dynamics & Optimization BDO'96*, St.Petersburg, July 1–5, 1995, St.Petersburg State University, will be published.

On the Mechanism of the Saw-Tooth Instability*

S. Heifets
Stanford Linear Accelerator Center, Stanford University, Stanford, CA
94309

Abstract

Dynamics of coherent modes of a single bunch above the threshold of microwave instability is discussed. The linear and nonlinear theories of bunch stability are presented and compared. It is shown, that normally the behavior of the system is the same as described by O'Neil [1] in plasma. Possible mechanisms of the saw-tooth instability are outlined.

1 INTRODUCTION

The microwave instability is one of the most intriguing phenomenon in the accelerator physics. The instability manifests itself in a seemingly simple situation of a single bunch under steady-state external conditions. At low current, a bunch can be described as N_b uncorrelated individual particles oscillating with the synchrotron frequency ω_{0s} in the one-dimensional time-independent external potential of the rf bucket. The particle motion may be described in terms of dimensionless canonical variables

$$x = z/\sigma_0, \quad p = -\delta/\delta_0, \quad \{x, p\} = 1 \qquad (1)$$

where z is position of a particle in respect with bunch centroid ($z > 0$ in the head of the bunch), $\delta = \Delta E/E$ is relative energy variation, σ_0 and δ_0 are the rms bunch length and the rms energy spread at zero current.

For small amplitudes $x \ll 1$, the rf potential is simply $U_{rf} = x^2/2$. At large N_b, dynamics substantially depends on the interaction of particles through the longitudinal wake fields W^δ excited in the beam pipe.

The wake $W^\delta(z_1 - z_2)$ generating by a particle at the location z_1 defines energy loss of a trailing particle, located at z_2, $\Delta E_l = -N_b e^2 W^\delta(z_l - z_t)$. The

---

*Work supported by Department of Energy contract DE–AC03–76SF00515.

wake does not depend on the bunch parameters being solely function of the geometry of the beam pipe and its conductivity. For the ultra-relativistic motion $\gamma = E/mc_0^2 \gg 1$, $W^\delta(z) = 0$ if $z < 0$. Although the wake field in the ultra-relativistic case is asymmetric, $W^\delta(z_1 - z_2) \neq W^\delta(z_2 - z_1)$, a single-particle dynamics still can be described as a motion in a self-consistent potential with Hamiltonian

$$H(x,p,s) = \frac{p^2}{2} + U(x,s), \quad \frac{\partial U(x,s)}{\partial x} = x - \lambda \sigma_0 \int dx' dp' \rho(x',p',s) W^\delta(\sigma_0(x'-x)). \tag{2}$$

Here the distribution function ρ is normalized by condition $\int dx' dp' \rho = 1$,

$$\lambda = \frac{N_b r_e}{2\pi R \gamma \alpha_E \delta_0^2}, \tag{3}$$

α_E is momentum compaction factor, and r_e is the classical electron radius. We use dimensionless time $s = \omega_{0s} t$. Note the relationship: $\omega_{0s}\sigma_0/c_0 = \alpha_E \delta_0$. Hence, the bunch stability depends only on two parameters, σ_0 and λ for a given wake field. The self-consistent Hamiltonian does not depend on time provided the single-particle distribution function is time-independent.

Synchrotron radiation introduces both damping of the single particle motion and quantum fluctuations exciting such a motion. Effect of the radiation damping and diffusion caused by quantum fluctuation can be described by the Fokker-Plank equation

$$\frac{\partial \rho}{\partial s} + \{H,\rho\}_{p,x} = \gamma_d \frac{\partial}{\partial p}[\frac{\partial \rho}{\partial p} + p\rho], \tag{4}$$

where γ_d is dimensionless radiation damping in units of ω_{0s} and curly brackets mean Pousson brackets. Eq. (4) has a steady-state solution, suggested by Haissinski [2]

$$\rho_H(x,p) = \frac{1}{Z_H} e^{-H_0(x,p)}, \tag{5}$$

generalizing Boltzmann's distribution to the system with self-consistent potential. Here $H_0(x,p)$ is given by Eq. (2) with $\rho = \rho_H$.

At the zero current, Haissinski solution gives just Gaussian distribution with temperature defined by quantum fluctuations (in our units, temperature $T = <p^2> = 1$). For finite N_b, explicit solution of the transcendental Eq. (5) can be obtained numerically starting with $\rho_H \propto \exp[-x^2/2]$ for large positive x and going backward in steps to large negative x. Haissinski solution, for reasonable wake fields, formally exist at any N_b. It describes quite well experimental data on the variation of the bunch profile with current [3] (so called potential well distortion, or PWD) and predicts that temperature (rms $<p^2>$) remains independent on N_b. However, the agreement breaks if the current exceeds certain threshold, where bunch become unstable. Instability, called the microwave instability, usually indicates itself by growing rms energy spread $<p^2>$ and, additional to the PWD, bunch lengthening.

It was noticed, that at least under certain conditions, instability can lead also to periodic variations of the bunch length. This kind of bunch oscillations was carefully studied in recent experiments in the damping ring at SLAC [4], but apparently was noticed as well in many laboratories before. In the initial observations, the rms bunch length, after relaxation of injection transients, changed periodically exponentially increasing and then also exponentially decreasing with typical time of the order of damping time. The variation looks like a relaxation oscillations with the pattern of σ_0 oscillations reminding saw-tooth. This gave the name of instability. The detail character of oscillations was sensitive to the rf voltage and current. With the onset of oscillations of the bunch length, a new line appeared in the bunch spectrum with frequency close but not equal to the frequency of the sextupole mode $3\omega_{0s}$. After the vacuum chamber of the ring was rebuild, the pattern of instability become more regular and the frequency of the instability has been changed to approximately $2\omega_{0s}$.

Instability in a system with Hamiltonian and damping is not trivial: usually, such a system goes with time to an equilibrium (although van-der-Pol oscillator gives an example of relaxation oscillations in the nonlinear system with damping). The cause of the instability is the asymmetry of the wake fields in the ultra-relativistic case. As the result, two-particle Hamiltonian is not Hermitian (action and contr-action are not equal). The situation is quite analogous to the system with radiation which is not Hermitian and does not conserve energy if radiation is not confined and, hence, probability of absorption of radiated quanta is negligible.

The spectrum of the saw-tooth instability shows that instability may be related to a single unstable coherent mode, while the microwave instability well above threshold may be attributed to excitation and chaotic interaction of many unstable modes simultaneously. If this is true, the saw-tooth instability provides a simplest model for the onset of the microwave instability. Study of the saw-tooth may give insight to the more complicated dynamics of the microwave instability itself.

Explanation of the saw-tooth instability requires study of the nonlinear dynamics of the system above the threshold of the microwave instability. A possible phenomenological explanation of the instability is based on the assumption that, above threshold of instability, a coherent mode becomes unstable, grows in time and starts changing the single particle distribution, stabilizing itself in this way. Then mode decays, its energy goes into uncorrelated single particle motion heating the bunch. After that, radiation damping cools system to the initial temperature and the process repeats again. Processes described here are known as the "overshoot phenomena". Another alternative for a system is to approach asymptotically a quasi-steady equilibrium with such a temperature (and the bunch length) that would adjust the threshold of instability to the bunch current.

Similar problem in plasma have been studied by O'Neil, and O'Neil and Morales [1] in 1965. In their theory, plasma oscillations with frequency Ω are described as a plasma wave with growing amplitude. The coherent wave traps particles and, as result, the coherent wave is stabilized. In O'Neil's theory, the growth rate of the wave oscillates in time and goes to zero after a few revolution periods T_m of trapped particles, see Fig. Consideration implied that damping is slow, $\Omega T_m \gg 1$.

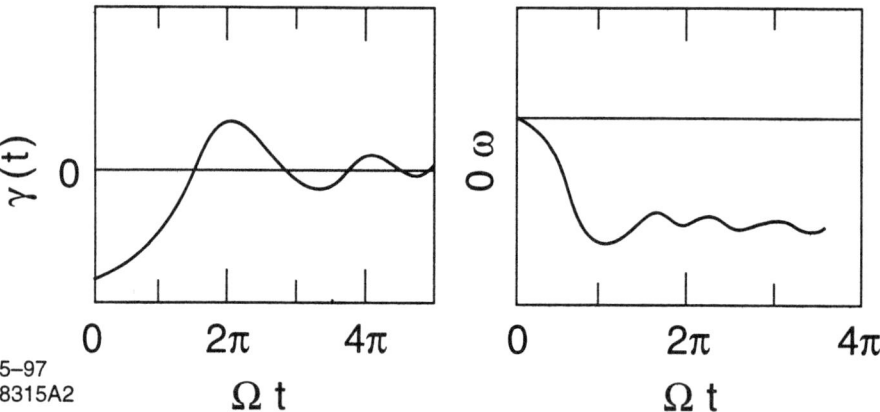

Figure 1: Dependence of the coherent frequency shift and the growth rate on time, from O'Neil and Morales [1].

In the accelerator physics study of instability is aimed mostly on finding the threshold of instability based on the linearized Vlasov equation. Yokoya and Chin [5] studied instability in the quasi-linear approximation, i.e. using the linearized Vlasov equation for non-zero azimuthal harmonics but retaining the averaged quadratic effect of the non-zero harmonic in the Fokker-Plank equation for ρ_0, which gave additional diffusion-like term. Results of their simulation show that the system above threshold approaches to a new quasi-steady equilibrium, maybe, after few oscillations.

Recently, Stupakov et. al. [6] in a different technique obtained for some range of parameters non-damping oscillations. However, interpretation of the results is not straightforward. The study is in progress.

Quite different approach was taken by Baartman and Dyachkov [7] who obtained relaxation-like oscillations of a bunch. Their model is based on the assumption that, initially, PWD generates potential well with two local minima with only one minimum being populated. Quantum fluctuations populates another minimum, and, as a result, potential well changes in such a way that two minima merge. Mixing and cooling return system to initial stage and process repeats. Although model esthetically is very attractive, the basic assumption requires usually too high current and hardly can be a universal mechanism of the saw-tooth instability.

In this paper we consider a possibility that initially unstable coherent mode approaches a steady-state equilibrium with a finite amplitude and zero damping without decaying into uncorrelated single particle motion. Such a mode can sustain itself by changing the single particle distribution accordingly and represents, essentially, a single non-linear Van-Kampen wave. First, we demonstrate how such a state can emerge in the perturbation theory, then obtain this state as a resonance solution in the nonlinear static theory, and, finally, discuss how system approach such a state in time.

2 PHENOMENOLOGY and PERTURBATION THEORY

Haissinski solution corresponds to the motion of particles in a self-consistent potential. For moderate currents, it can be presented as a polynomial expansion around the equilibrium x_0:

$$U_H(x) = \frac{\hat{\omega}_s^2 (x-x_0)^2}{2} + \alpha \frac{(x-x_0)^3}{3} + \beta \frac{(x-x_0)^4}{4} + ..., \qquad (6)$$

where x_0, the nonlinear coefficients, and synchrotron frequency $\hat{\omega}_s$ depend on current.

Trajectory of a particle in such a potential is superposition of harmonics

$$x(J,\phi) = A[1 - \frac{\alpha^2 A^2}{18\omega_{0s}^4}]\sin\phi - \frac{\alpha A^2}{2\omega_{0s}^2}[1 + \frac{1}{3}\cos 2\phi] - \frac{A^3}{16\omega_{0s}^2}[\frac{\alpha^2}{3\omega_{0s}^2} + \frac{\beta}{2}]\sin 3\phi + ... \qquad (7)$$

where $d\phi/ds = \hat{\omega}_s$.

It is convenient to use angle-action variables ϕ, J, where the action $J = \int (dx/2\pi)\sqrt{[2(E - U_H(x)]}$. In this variables, ρ_H and the Hamiltonian H_0 are independent of ϕ:

$$H_0 = H_0(J), \quad \rho_H = \frac{1}{Z_H}e^{-H_0(J)}. \qquad (8)$$

The current dependent synchrotron frequency $\omega_s(J)$ is given by $d\phi/ds = \omega_s(J) = \frac{\partial H_0}{\partial J}$, and the amplitude A in Eq. (7) is $A = \sqrt{2J/\omega(J)}$.

To study stability of Haissinski solution, introduce perturbation u

$$H_0(J) \to H(\phi, J, s) = H_0(J) + u(J, \phi, s). \tag{9}$$

The Fokker-Plank equation for the distribution function $\rho(J, \phi, s)$ takes form

$$\frac{\partial \rho}{\partial s} + \{H_0 + u, \rho\}_{J,\phi} = \hat{D}\rho \tag{10}$$

where the right-hand-side (RHS), averaged over phase ϕ, is [8]

$$\hat{D}\rho = \gamma_d \frac{\partial}{\partial J}[\frac{J}{\omega(J)}\frac{\partial \rho}{\partial J} + J\rho]. \tag{11}$$

Expand ρ in azimuthal harmonics

$$\rho(J, \phi, s) = \sum_m \rho_m(J, s)e^{im\phi}, \qquad \rho_n(J, s) = \rho^*_{-n}(J, s). \tag{12}$$

and take similar expansion for the perturbation u.

Equation for the azimuthal harmonics $\rho_m(J, s)$ can be obtained from the Fokker-Plank equation Eq. (10). The RHS describes slow processes of diffusion and damping and can be, in this case, neglected. In the perturbation theory, u is a small parameter. Neglecting interaction of the non-zero azimuthal harmonics $\rho_n \propto u$, we get

$$\frac{\partial \rho_n(J, s)}{\partial s} + in\omega_{eff}(J)\rho_n - in\frac{\partial \rho_0(J, s)}{\partial J}u_n(J, s) = 0, \tag{13}$$

where $\omega_{eff} = \omega_s(J) + du_0/dJ$.

Eq. (13) can be solved assuming that ρ_0 depends on time adiabatically while harmonics of the perturbation u_n oscillate in time. Harmonics $\rho_n(J, s)$ depend on time in the same way,

$$u_n(J, s) = \hat{u}_n(J)e^{-in\Omega s}, \qquad \rho_n(J, s) = \hat{\rho}_n(J)e^{-in\Omega s}, \tag{14}$$

where $\Omega = \Omega_0 + i\Gamma$. This can be generalized replacing $\Omega s \to \int^s ds' \Omega(s')$ allowing adiabatic variation of Ω with time.

Perturbation u is generated by the perturbation of the distribution function ρ. They are related by the condition of self-consistency:

$$u(J, \phi, s) = \lambda \sigma_0 \int dJ' d\phi'[\rho(J', \phi', s) - \rho_H(J', \phi')]S[x(J', \phi') - x(J, \phi)], \tag{15}$$

where $S(x) = \sigma_0 \int_0^x dx' W^\delta(\sigma_0 x')$. In terms of the impedance $Z(\omega)$, defined as Fourier transform of the wake,

$$S(x) = \frac{4\pi}{Z_0}\int \frac{d\omega}{2\pi i}\frac{Z(\omega)}{\omega}[1 - e^{-i(\omega\sigma_0/c_0)x}]. \tag{16}$$

Here $Z_0 = 4\pi/c_0 = 120\pi$ Ohm is impedance of vacuum.

With definitions of Eqs. (12),(16), Eq. (15) can be written as

$$u_m(J,s) = -2\pi\lambda \int dJ' R_m(J,J') \rho_m(J',s). \tag{17}$$

Here, u_m and ρ_m can be replaced by the amplitudes $\hat{u}_m$ and $\hat{\rho}_m$ respectively.

The kernel

$$R_m(J,J') = \frac{4\pi}{Z_0} \int \frac{d\omega}{2\pi i} \frac{Z(\omega)}{\omega} C_m(J,\omega) C^*(J';\omega), \tag{18}$$

where

$$C_m(J,\omega) = \int \frac{d\phi}{2\pi} e^{-im\phi} e^{ix(J,\phi)}. \tag{19}$$

In Eq. (17) we averaged out fast oscillating terms proportional to $e^{ik\omega_s s}$ with $k \neq 0$. In other respects, it is exact equation.

For small nonlinearities, coefficients C_m can be calculated using trajectory Eq. (7)

$$C_m(J,\omega) = J_m(a) - i\alpha J \frac{\omega\sigma_0}{c_0\omega_s^3}[J_m(a) + \frac{1}{6}J_{m-2}(a) + \frac{1}{6}J_{m+2}(a)] + ..O(\alpha^2,\beta), \tag{20}$$

where $a = (\omega\sigma_0/c_0)\sqrt{2J/\omega_s}$.

Eqs. (13) and (17) give

$$\frac{\partial \rho_n(J,s)}{\partial s} + in\omega_{eff}(J)\rho_n = -2in\pi\lambda \frac{\partial \rho_0(J,s)}{\partial J} \int dJ' R_n(J,J')\rho_n(J',s). \tag{21}$$

For the amplitude $\hat{\rho}_n$, see Eq. (14), we get

$$[\omega_{eff}(J) - \Omega]\hat{\rho}_n = -2\pi\lambda \frac{\partial \rho_0(J,s)}{\partial J} \int dJ' R_n(J,J')\hat{\rho}_n(J',s). \tag{22}$$

Eq. (22) has a solution

$$\hat{\rho}_n(J) = \kappa_n \delta(J - J_r) + \frac{f_n(J)}{\omega_{eff}(J) - \Omega} \frac{\partial \rho_0(J,s)}{\partial J}, \tag{23}$$

with arbitrary κ_n, and J_r defined by $\omega_{eff}(J_r) = \Omega$. The new function $f_n(J)$ is defined by equation

$$f_n(J) = -2\pi\lambda[\kappa_n R_n(J,J_r) + \int dJ' R_n(J,J') \frac{\partial \rho_0(J,s)}{\partial J} \frac{f_n(J')}{\omega_{eff} - \Omega}]. \tag{24}$$

Comparison of Eqs. (17) and (24) gives $f_n = \hat{u}_n(J)$.

Eq. (23) describes Van-Kampen wave [9] which, in plasma physics, corresponds to a jet of particles plus distorted distribution of the background plasma proportional to f_n and localized around resonance amplitude J_r. Van-Kampen solutions are stationary waves with real frequency Ω. Solution of the initial value problem according to Landau [10] is given by homogeneous Eq. (24) with complex eigen-value Ω. The homogeneous equation corresponds to the limit $\kappa_n \to 0$ and infinitesimal shift of $\Omega \to \Omega + i\epsilon$, $\epsilon > 0$, what gives the dispersion equation of the linear theory

$$f_m(J) = -2\pi\lambda \int dJ' R_m(J,J') \frac{\partial \rho_0(J')}{\partial J'} \frac{f_m(J')}{\omega_{eff}(J') - \Omega - i\epsilon}. \tag{25}$$

Eq. (25) defines the eigen value $\Omega = \Omega_0 + i\Gamma$ and, for $\Gamma < 0$ Eq. (14) describes decay of the initial state into stationary Van-Kampen waves what usually is referred to as Landau damping.

For the zero harmonic, the Fokker-Plank equation gives

$$\frac{\partial \rho_0}{\partial s} + \sum_m im \frac{\partial}{\partial J}[\rho_m u_m^*] = \gamma_d \frac{\partial}{\partial J}[\frac{J}{\omega_s(J)} \frac{\partial \rho_0}{\partial J} + J\rho_0] \tag{26}$$

Substitute Landau solution for ρ_m from Eq.(23) with $\kappa_n = 0$ and average over the fast oscillations. Then,

$$\frac{\partial \rho_0}{\partial s} = \gamma_d \frac{\partial}{\partial J}\left[\frac{J}{\omega_s(J)} \frac{\partial \rho_0}{\partial J} + J\rho_0 + \sum_{m>0} \frac{2m}{\gamma_d}|f_m(J)|^2 \frac{\Gamma e^{2m\Gamma s}}{(\omega_{eff} - \Omega_0)^2 + \Gamma^2} \frac{\partial \rho_0}{\partial J}\right]. \tag{27}$$

The last term describes tha averaged effect of the unstable mode on $\rho_0(J)$. For Van-Kampen waves the last term in the brackets of the RHS should be replaced by

$$\frac{2m}{\gamma_d} Im[\kappa_m f_m^*]\delta(J - J_r). \tag{28}$$

This expression can be obtained from the last term of Eq. (27) in the limit $\Gamma \to 0$ and $\pi f_m(\partial \rho_0/\partial J) \to \kappa_m$.

Solution of Eq. (27) can be obtained in a quasi-steady-state approximation assuming that variation of ρ_0 in time, $(1/\gamma_d)(\partial \rho_0/\partial s) \propto (\Gamma/\gamma_d)\rho_0$ is slow compared to the spatial variation $\partial \rho_0/\partial J \propto \rho_0/\Delta J$, where ΔJ is the range of J where ρ_0 is significantly different from ρ_H. Usually $\Gamma/\gamma_d \simeq 1$, and this condition means that $\Delta J << 1$. In this approximation, we can neglect the LHS, and get

$$\frac{J}{\omega_s(J)} \frac{\partial \rho_0}{\partial J}[1 + \Phi(J,s)] + J\rho_0 = 0, \tag{29}$$

where

$$\Phi(J,s) = \frac{\omega_s(J)}{\gamma_d J} \sum_{m>0} 2m|f_m(J)|^2 \frac{\Gamma}{(\omega_{eff} - \Omega_0)^2 + \Gamma^2} e^{2m\Gamma s}. \tag{30}$$

Hence,
$$\rho_0 = \frac{1}{Z} e^{-\int \frac{\omega_s(J)dJ}{1+\Phi(J,s)}}, \tag{31}$$

For $\Phi = 0$, Eq. (31) gives, of course, $\rho_0 = \rho_H$.

To understand effect of $\Phi(J,s)$, let us assume that particles can be in resonance $\omega(J_r) = \Omega$ only for the n-th azimuthal mode, and approximate

$$\Phi(J,s) = \frac{\Phi_0}{(J-J_r)^2 + \Delta^2} \tag{32}$$

where $\Delta = |\Gamma/(d\omega_{eff}/dJ)|$, and

$$\Phi_0 \simeq \frac{2n\omega_s}{J_r} \frac{\Gamma}{\gamma_d} \left|\frac{f_n}{d\omega_{eff}/dJ}\right|^2_{J=J_r} e^{2n\Gamma s} \tag{33}$$

In this approximation, the exponent in Eq. (31)

$$\int_0^J \frac{dJ}{1+\Phi(J,s)} \simeq J - \frac{\Phi_0}{\sqrt{\Delta^2+\Phi_0}} \arctan[\frac{J-J_r}{\sqrt{\Delta^2+\Phi_0}}]. \tag{34}$$

Eq. (34) illustrates the main result of the quasi-linear theory: the number of particles increases at $J > J_r$ and decrease at $J < J_r$ while there is transition around $J = J_r$ with the width $J - J_r = \pm\sqrt{\Delta^2+\Phi_0}$ where the distribution function is, essentially, flat. The width of the transition $\Delta\omega = |\omega'(J-J_r)|$ initially is of the order of Γ, while, for large Φ_0, the width

$$\Delta\omega = |f_n|\sqrt{\frac{2n\omega_s\Gamma}{J_r\gamma_d}} e^{n\Gamma s}. \tag{35}$$

Eqs. (25) and (34) suggest a mechanism of the saw-tooth relaxation oscillations. The Haissinski distribution may give an unstable solution of Eq. (25) for some mode with $\Gamma > 0$ localized around J_r. The growing mode changes distribution function from ρ_H to ρ_0, Eq. (31), which becomes increasingly flat with time around the resonance amplitude J_r. Hence, the kernel of Eq. (25), proportional to $\partial\rho_0/\partial J$, decreases in time what can stabilize growth or, in some cases, may change sign of Γ. In the last case, the growing mode decay, radiation cooling takes the system back to Haissinski distribution, and process repeats. If, on the other hand, $\Gamma \to 0$, there is a possibility that the system goes to a steady-state regime with finite amplitude of periodically oscillating mode and ρ_0 having zero derivative at $J = J_r$.

Note, that for a Van-Kampen wave with periodically oscillating azimuthal mode ρ_m, Eq. (29) takes form

$$\frac{J}{\omega_s(J)}\frac{\partial\rho_0}{\partial J} + J\rho_0 + \frac{2m}{\gamma_d}Im[\kappa_m f_m^*]\delta(J-J_r) = 0, \tag{36}$$

and gives a steady-state distribution ρ_0 with a kink at $J = J_r$.

3 RESONANCE SOLUTION

The solution obtained in the previous section describes self-consistent oscillating azimuthal mode with ρ_0 being different from Haissinski solution. The solution is similar to the Van-Kampen wave but does not have a discontinuity at $J = J_r$. To clarify the structure of such a solution, we restrict ourself to a periodic perturbation corresponding to a single n-th azimuthal mode. In this case, results of the perturbation theory can be improved. Following Schonfeld [8] and Meller [11], the Hamiltonian of Eqs.(9),(14)

$$H(J,\phi,s) = H_0(J) + \epsilon_n \cos(n\phi - \int^s \Omega ds + arg(f_n)), \quad \epsilon_n = 2|f_n|, \quad (37)$$

can be made time-independent by changing variables from J, ϕ to $j = J - J_r$, and $\alpha = \phi - \int^s \frac{\Omega}{n} ds + (\pi + arg(f_n))/n$, i.e. describing system in the frame rotating in the phase space with frequency $\omega_r = \omega(J_r) = \Omega/n$. If Ω varies slowly with time, than $\dot{J}_r = dJ_r/ds \neq 0$. In these variables, the Hamiltonian Eq. (37) is the Hamiltonian of a nonlinear pendulum

$$H(\alpha,j) = H_0(J_r+j) - \frac{\Omega}{n}(J_r+j) + \epsilon_n(1 - \cos n\alpha) \simeq \frac{j^2}{2M} + \epsilon(1 - \cos n\alpha) + \dot{J}_r\alpha, \quad (38)$$

where $(1/M) = \omega_r' = (d\omega_s(J)/dJ)_{J=J_r}$. It is convenient to use parameter κ related to particle energy $E \equiv 2\epsilon/\kappa^2$. Then, particles with $\kappa > 1$ are in the resonance with the perturbation, they are trapped and their motion is bounded. Unbounded particles have $\kappa < 1$ and momentum

$$j = \pm\sqrt{\frac{4M\epsilon}{\kappa^2}[1 - \kappa^2 \sin^2(n\alpha/2)]}. \quad (39)$$

with the sign corresponding to particles above and below the fixed points $j = 0$, $\alpha = 2\pi m/n$, $m = 0, 1, .., n-1$. For the trapped particles $\kappa > 1$, their trajectories are enclosed by the separatrices $\kappa = 1$.

For the n-th harmonics there are n fixed points in the phase plane. It is more convenient to change variables to canonical $q = 2j/n$ and $\beta = n\alpha/2$. The full range of variation of β is $n\pi$, what correspond to the variation of α within 2π. We will consider the range of β-s within one π, changing the norm of the distribution function to $\int_\pi \rho dq d\beta = 1$.

Introduce now canonical variables r, ψ to make Hamiltonian Eq. (38) independent on ψ. For the unbounded motion, the action

$$r = \int_\pi \frac{d\beta}{\pi} q = \frac{8\sqrt{M\epsilon}}{\pi n} \frac{E(\kappa)}{\kappa}, \quad (40)$$

and the generating function

$$G(r,\beta) = \pm \frac{4\sqrt{\epsilon M}}{n\kappa(r)} E(\beta, \kappa(r)). \quad (41)$$

Here E is elliptic integral defined in Gradshteyn [12]. The phase and the frequency $\omega_M = dH(r)/dr$, $H = 2\epsilon/\kappa^2$, are correspondingly

$$\psi = \frac{\partial G(r,\beta)}{\partial r} = \pm\frac{\pi}{2K(\kappa)}F(\beta,\kappa), \qquad \omega_M = \frac{\pi n}{2\kappa K(\kappa)}\sqrt{\frac{\epsilon}{M}}. \qquad (42)$$

The Fokker-Plank Eq. (10) in this variables takes form

$$\frac{\partial \rho}{\partial s} + \{H,\rho\}_{r,\psi} = \gamma_d[\frac{\partial}{\partial r}L_1\rho - \frac{\partial}{\partial \psi}L_2\rho] - \dot{J}_r\{\alpha,\rho\}, \qquad (43)$$

where operators in the RHS are

$$L_1 = [(\frac{\partial x}{\partial \psi})^2 \frac{\partial}{\partial r} - (\frac{\partial x}{\partial \psi}\frac{\partial x}{\partial r})\frac{\partial}{\partial \psi} + p], \qquad (44)$$

$$L_2 = [(\frac{\partial x}{\partial \psi}\frac{\partial x}{\partial r})\frac{\partial}{\partial r} - (\frac{\partial x}{\partial r})^2 \frac{\partial}{\partial \psi} + p]. \qquad (45)$$

The momentum $p(r,\psi)$ is given by

$$p = \{x, H(x,p)\}_{x,p} = \{x, H(r) + \omega_r j\} = \omega_M \frac{\partial x}{\partial \psi} + \omega_r \frac{\partial x}{\partial \alpha}. \qquad (46)$$

Expand

$$\rho(r,\psi) = \sum_k \rho_k(r,s)e^{ik\psi}. \qquad (47)$$

In the RHS of equations for harmonics $\rho_k(r,s)$ we keep only terms proportional to $\rho_0(r,s)$. After some transformations, the Fokker-Plank equations for $\rho_0(r,s)$ and non-zero harmonics ρ_k take form:

$$\frac{\partial \rho_0}{\partial s} = \gamma_d \frac{\partial}{\partial r}[a_0(\frac{\partial \rho_0}{\partial r} + \omega_M \rho_0) + \omega_r b_0(1 - \frac{\dot{J}_r}{\gamma_d J_r})\rho_0], \qquad (48)$$

$$\frac{\partial \rho_k}{\partial s} + ik\omega_M \rho_k = \gamma_d \frac{\partial}{\partial r}[a_k \frac{\partial \rho_0}{\partial r}] + \gamma_d(a_k \omega_M + \omega_r b_k)\frac{\partial \rho_0}{\partial r} \qquad (49)$$

$$-\gamma_d \frac{1}{\omega_M}(\frac{\partial(\omega_M a_k)}{\partial r})\frac{\partial \rho_0}{\partial r} + \dot{J}_r \frac{\partial \rho_0}{\partial r} <e^{-ik\psi}\frac{\partial \alpha}{\partial \psi}>. \qquad (50)$$

Here

$$a_k = <(\frac{\partial x}{\partial \psi})^2 e^{-ik\psi}>, \quad b_k = <(\frac{\partial x}{\partial \psi})(\frac{\partial x}{\partial \alpha})e^{-ik\psi}> \qquad (51)$$

and $< .. >$ means averaging over ψ. Note $<\frac{\partial \alpha}{\partial \psi}> \neq 0$ for unbounded particles.

In the steady-state, Eq.(49) has a solution [11]

$$\rho_0(r) = \frac{1}{Z_M} \exp -[H(r) + \sigma(r)], \qquad (52)$$

where

$$\sigma(r) = \pm 2\omega_r \sqrt{\epsilon M} \theta(1-\kappa)[\frac{1}{\kappa} - \Psi(\kappa)], \quad \Psi(\kappa) = 1 - \int_\kappa^1 \frac{du}{u^2}[\frac{\pi}{2E(u)} - 1]. \qquad (53)$$

The steady-state resonance solution Eq. (52) in J, ϕ variables is a sum of $\rho_0(J)$ and non-zero harmonics oscillating in time as $e^{in(\phi-\Omega s)}$ with real Ω, similarly to the result obtained in the perturbation theory. The zero-th harmonic describes modification of the Haissinski solution $\rho_H(J)$ by trapped particles. As in the perturbation theory, it is different from ρ_H by a factor $\exp[\pm 2\omega_r \sqrt{\epsilon M} \Psi(0)]$ at large distances from the resonance $J = J_r$ being approximately constant within the separatrix $\kappa > 1$, where trapped particles move with frequency $\omega_M \simeq \sqrt{4\epsilon/M}$. Unlikely to the perturbation theory, the width of the resonance $|J - J_r| = \sqrt{4M\epsilon}$, or $\Delta\omega(J) \simeq \omega_M$, and is proportional to $\sqrt{f_n}$ rather than f_n as in Eq. (35). It is worth noting, that although the resonance solution does not describe any turbulence, the rms energy spread in this state is different from the rms induced by the synchrotron radiation, which remains unchanged in the Haissinski solution.

Both Haissinski and the resonance solutions satisfy the Fokker-Plank equation. To choose one of them, the free energy of the solutions have to be compared. It was shown [11] [13] that for some parameters the resonance solution, indeed, may have lower free energy and becomes preferable meaning that Haissinski solution becomes globally unstable.

4 TIME DEPENDENCE

So far the amplitude of the perturbation ϵ was arbitrary. Condition of self-consistency Eq. (17) defines its value.

Eq. (50) for the non-zero harmonics $k \neq 0$ can be simplified using solution Eq.(52):

$$\frac{\partial \rho_k}{\partial s} + ik\omega_M \rho_k = \gamma_d \left\{ \omega_r b_k (1 - \frac{J_r}{\gamma_d J_r}) \frac{\partial \rho_0}{\partial r} - \omega_M a_k \frac{\partial}{\partial r}[\frac{1}{\omega_M} \frac{\partial \sigma}{\partial r} \rho_0] \right\}. \qquad (54)$$

Solution of this equation with zero initial conditions is

$$\rho_k(r,s) = \gamma_d \int_0^s ds' e^{-ik\omega_M(s-s')} \left\{ \omega_r b_k (1 - \frac{J_r}{\gamma_d J_r}) \frac{\partial \rho_0}{\partial r} - \omega_M a_k \frac{\partial}{\partial r}[\frac{1}{\omega_M} \frac{\partial \sigma}{\partial r} \rho_0] \right\}. \qquad (55)$$

Substitute this into Eq. (17) which can be written as

$$\epsilon_m(J,s) = -\lambda \int dJ' d\phi' R_m(J,J') e^{-im\phi'} \rho(J',\phi',s). \quad (56)$$

Change variables in the integral to r, ψ, substitute $\rho = \rho(r,\psi,s)$ expanded in harmonics Eq. (47), and use Eq.(55). After summation over harmonics

$$\rho(r,\psi,s) = \rho_0(r,s) + \gamma_d \int_0^s ds' [\Psi(r,\psi-\omega_M(s-s'),s') - <\Psi(r,\psi,s')>] \quad (57)$$

where

$$\Psi(r,\psi,s) = J_r(1 - \frac{\dot{J}_r}{\gamma_d J_r})(\frac{\partial \beta(r,\psi)}{\partial \psi})\frac{\partial \rho_0(r,s)}{\partial r} - J_r \frac{\omega_M}{\omega_r}(\frac{\partial \beta}{\partial \psi})^2 \frac{\partial}{\partial r}[\frac{\rho_0}{\omega_M} \frac{\partial \sigma(r,s)}{\partial r}]. \quad (58)$$

Function $\Psi(r,\psi,s)$ is periodic in ψ, what follows from periodicity of the elliptic integral $F(\beta + n\pi, \kappa) = F(\beta, \kappa) + 2nK(\kappa)$, and can be expanded in series

$$\Psi(r,\psi,s) = \sum_l \Psi_l(r,s) e^{2il\psi}. \quad (59)$$

The integration over ds' in Eq. (57) can be carried out:

$$\rho(r,\psi,s) = \rho_0(r,s) + \gamma_d \sum_{l \neq 0} \frac{\Psi_l(r)}{2il\omega_M(r)}(1 - e^{-2il\omega_M s}). \quad (60)$$

Hence, $\rho(r,\psi,s)$ is periodic in time s with period $2\pi/\omega_M$.

Substitute Eq. (60) in Eq.(56). Contribution of ρ_0 at $\epsilon = 0$ has been already included into definition of the unperturbed self-consistent Hamiltonian $H(r)$. The difference $\rho_0^\epsilon - \rho_0^{\epsilon=0}$ depends on time through dependence of $\epsilon(s)$. Therefore, it changes a coefficient in the LHS and, for $\lambda << 1$, can be neglected. Hence,

$$\frac{d\epsilon_n}{ds} = -\gamma_d \lambda \sum_{l \neq 0} \int dr e^{-il\omega_M(r)s} \Phi_l(r,s), \quad (61)$$

where

$$\Phi_l(r,s) = \Psi_l(r,s) \int d\psi R_n(J_r, J_r) e^{-in\alpha(r,\psi)+il\psi}. \quad (62)$$

At large s the main contribution is given by the vicinity of the separatrix $\kappa \simeq 1$. The area within the separatrix gives zero contribution, because, in this case, $\sigma(r) = 0$ and the first term in Ψ, Eq. (58), has opposite signs above and below the fixed point. Integral in Eq. (62) is of the order of $\pi/\sqrt{2lK(\kappa)}$, and Ψ_l for unbounded motion can be expressed in series using expansion of elliptical functions. For our purpose it is suffice to note that, choosing ω_M as a new variable, we can estimate the integral

$$\frac{d\epsilon_n}{ds} \propto \int dr \Phi_l(r) e^{-il\omega_M s} = \sqrt{\frac{\epsilon}{M}} \int \frac{d\omega_m}{\omega_m^3} e^{-\frac{\pi n}{\omega_M}\sqrt{\epsilon/M} - il\omega_M s} \Psi_l(\omega_M) \propto e^{-(1+i)\sqrt{\zeta s}}, \tag{63}$$

where $\zeta = (\pi n l/2)\sqrt{\epsilon/M}$.

Hence, ϵ behaves in time as damped oscillator due to frequency spread of particles in the vicinity of the separatrix. The estimate has to be confirmed by numerical calculations. This result is in agreement with result obtained by O'Neil and Morales [1]. We can expect, therefore, that the system approaches a steady-state described by a resonance solution provided, of course, that its free energy lower than that of Haissinski solution. Behavior at small s can be defined from the perturbation theory and describes the growing unstable mode.

5 Conclusion

Dynamics of the system in the nonlinear regime above the threshold of instability may be quite complicate. We tried to present arguments in favor of the scenario where the unstable mode may relax to a resonance solution representing a nonlinear Van-Kampen wave in the phase space of a bunch. Such a state implies a self-consistent solution where the single particle distribution is changed to support a stable coherent mode with finite amplitude. From the experimental point of view, spectrum of a bunch in such a state would be different from the spectrum of Haissinski bunch, with corresponding change of the rms bunch length *and* of the energy spread. However, it is possible that several mechanisms of the instability can co-exist or be preferable for different range of parameters, including Dyachkov-Baartman mechanism or over-shoot phenomenon. The instability can be stabilized when the separatrix grows to cover substantial part of the finite phase space of the bunch. If there are two unstable modes growing at the same time, they may start to interact with each other when their separatrices start overlapping what would generate chaotic layers and, eventually, destroy the separatrices. It is possible also, that relaxation oscillations can be induced if the derivative of the synchrotron frequency $\omega' = (d\omega_s(J)/dJ)_{J=J_r}$ changes sign for trapped particles while the potential well is changed by the growing unstable mode. Indeed, the sign of ω' defines which point in the phase space is stable (elliptic) and which is unstable (hyperbolic). Particles trapped in the separatrix of a stable point become free when ω' changes sign and may be re-captured by the new stable (former-unstable) point, generating completely new conditions for stability of the coherent modes. This mechanism of relaxation oscillations maybe related to the instability found in linearized theory by Oide [14].

Microwave instability still suggest more questions than we are able to answer today. More efforts including new ideas and numeric analysis are needed to understand and explore all possibilities.

References

[1] O'Neil, T. Phys. Fluids 1965, 8, pp. 2255.

Morales,G.T. and O'Neil,T. Phys. Rev. 1965, 28, pp. 417.

Figure 2 is taken from the article by Oraevsky, V.N Kinetic theory of waves, in Basic Plasma Physics I, Handbook of Plasma Physics Volume 1, Editors Galeev, A.A, and. Sudan, R.N, North-Holand Publish. Co., 1983.

[2] Haissinski, J. Nuovo Cimento, 1973, 18B, pp. 72.

[3] Bane, K. "Bunch Lengthening in the SLC Damping Ring", 1990, SLAC-PUB-5177.

[4] Krejcik, P. et al., Proc. IEEE Part. Accel. Conf., Washington D.C., 1993, p. 3240.

[5] Chin Y., and Yokoya, K., Nucl Instr. and Methods, 1984, 226, 223-249.

[6] Stupakov, G.V.,. Breizman, B.N, and Pekker, M.S., SLAC-PUB-7377, 1996

[7] Baartman, R., and Dyachkov, M. Proc. IEEE Part. Accel. Conf., Dallas, 1995.

[8] Schonfeld, J., Ann. Phys., 1985, 160, pp. 149.

[9] Van Kampen, N.G., Physica 1955, 21, pp. 949.

[10] Landau, L.D., Zh. Eksp. Teor. Fiz., 1946, 16, pp. 574.

[11] Meller, R.E., Ph. D. Thesis, "Statistical Method for Nonequilibrium Systems with Application to Accelerator Beam Dynamics", Cornell, 1986.

[12] Gradshteyn, I.S., and Ryzhik, I.M., "Table of Integrals, Series, and Products", Academic Press, New York, 1980.

[13] Heifets, S.A., Phys. Rev. 1996, 54, p.2889.

[14] Oide, K., "A New Mechanism of Longitudinal Single Bunch Instability in Storage Ring", KEK, 1995 (unpublished).

Integrable Cases In Nonlinear Betatron Motion

Johannes Hagel

Universidade da Madeira, Portugal

Abstract. The integrability of the one dimensional betatron equation of motion is discussed . Although it is known that the general time dependent differential equation of second order describing single particle motion in presence of sextupoles is nonintegrable, sp ecial cases may be found in which integrability can be proven and first integrals can be written analytically. The present paper introduces a method to find specific sextupole distributions for which integrability can be obtained. The solutions of such in tegrable equations are investigated numerically and analytically and a rigorous stability analysis of the solutions with respect to their initial conditions is performed. The possibilities of applications of this theory to real machines is discussed.

I INTRODUCTION

The equations of betatron motion in a circular accelerator or storage ring containing higher order multipoles are nonlinear and explicitly time dependent. These equations, being derivable from a Hamiltonian are generically nonintegrable, meaning that no a lgebraic or even transcendental analytic function of the phase coordinates and the time can be established that is constant along the particle trajectories for any set of initial conditions. One important implication of this statement is that we are not a ble to find general restrictions of the particles motion to invariant surfaces in phase space. This fact complicates considerably the analytic (and numeric) search for boundaries of stability in the space of initial conditions and makes the search for the maximum stable oscillation amplitude of betatron motion (frequently called the dynamical apert ure) very difficult. In this paper we investigate a case of a purely horizontal motion of particles in a storage ring with some sextupole distribution along the circumference. The equation of motion for this case is (after the well known Courant and Snyder transformation, Ref . Courant and Snyder, 1952)

$$\frac{d^2x}{d\theta^2} + Q^2 x + \frac{1}{2} Q^2 \beta^{5/2}(\theta) K_3(\theta) x^2 = 0 \qquad (1)$$

where β is the horizontal focusing function, Q the horizontal tune and K_3 the sextupole distribution along the ring. The angle θ is related to the distance s by the usual Courant and Snyder transformation

$$\theta = \frac{1}{Q} \int_0^s \frac{ds}{\beta(s)} \qquad (2)$$

This equation being nonlinear and explicitly depending on the independent variable is generally nonintegrable. However, one may try to find specific sextupole distributions that render the equation integrable despite of its time dependence. A method to obtain such special coefficient functions for precisely this type of equations of betatron motion has been developed by [2]. It is shown that a function depending on five parameters can be found which produces the desired integrability and in this way enables us to derive a straightforward stability analysis of the nonlinear equation.

In the second section the equation is introduced and integrability is defined. In section 3 the method for finding integrable cases is explained and all possible coefficient functions leading to first integrals being quadratic polynomials in the canonical momentum are derived. However, it turns out that two of the five parameters mentioned have to be set to zero in order to solve the defining equations for the polynomial coefficients in closed form. In the fourth section we numerically and analytically in vestigate the properties of the solutions of the found integrable cases. It is shown that the solutions of these integrable , time dependent equations behave in a rather complex way and can be well compared to the generic behaviour of particle trajectorie s in accelerators and storage rings. On the other hand, finding the analytical stability limit w.r.t. the initial conditions turns out to be reducible to solving a simple cubic equation and is done fully explicitly. Section 5 discusses briefly the possibl e applications to accelerator physics.

II TIME DEPENDENT EQUATION

We consider the second order, nonlinear and explicitly time dependent equation of motion

$$\ddot{x} + \omega^2 x + f(t) x^2 = 0 \qquad (3)$$

where ω is a real number and $f(t)$ a real function of time. The Hamiltonian function of this equation is given by ($p = \dot{x}$)

$$H(x, p, t) = \frac{1}{2}(\omega^2 x^2 + p^2) + \frac{1}{3} f(t) x^3 \qquad (4)$$

$$\dot{x} = \frac{\partial H}{\partial p} = p \tag{5}$$

$$\dot{p} = -\frac{\partial H}{\partial x} = -\omega^2 x - f(t)x^2 \tag{6}$$

and is NO integral of this equation since $\partial H/\partial t \neq 0$ [3]. However this does not exclude that some function of x, p and the time t may exist that fulfills the condition

$$\frac{dI(x,p,t)}{dt} = 0 \tag{7}$$

and therefore represents a constant of motion of the dynamical system described by Eq. (3). In such a case the dynamical system related to Eq. (3) is integrable (since the problem is one dimensional) and all solutions will lie on either closed or open invariant surfaces in the three dimensional space (x, p, t).

III POLYNOMIAL TRIAL FUNCTION AND FIRST INTEGRAL

We look for a constant of motion of Eq. (3) in a set of functions defined by

$$I(x,p,t) = \alpha_0(q,t) + \alpha_1(q,t)p + \alpha_2(q,t)p^2 \tag{8}$$

where the $\alpha_i(q,t)$ are yet unknown functions. If we succeed defining these functions in a unique way by inserting the trial function (8) into the condition (7) we have indeed found a first integral of Eq. (3). Proceeding in this way and using Hamiltons equations written in the previous section we obtain the following condition to be satisfied in order to obtain integrability:

$$p\left[\frac{\partial \alpha_0}{\partial x} + \frac{\partial \alpha_1}{\partial x}p + \frac{\partial \alpha_2}{\partial x}p^2\right] - (\omega^2 x + f(t)x^2)(\alpha_1 + 2\alpha_2 p) + \frac{\partial \alpha_0}{\partial t} + p\frac{\partial \alpha_1}{\partial t} + p^2\frac{\partial \alpha_2}{\partial t} = 0 \tag{9}$$

This polynomial expression in the canonical momentum p can only be identical to zero if all the coefficients meet this requirement. This leads us to the following recursive linear system of linear partial differential equations.

$$p^3 : \quad \frac{\partial \alpha_2}{\partial x} = 0 \tag{10}$$

$$p^2 : \quad \frac{\partial \alpha_2}{\partial t} + \frac{\partial \alpha_1}{\partial x} = 0 \tag{11}$$

$$p^1 : \quad \frac{\partial \alpha_1}{\partial t} - 2\alpha_2 f(t)x^2 - 2\alpha_2 \omega^2 x + \frac{\partial \alpha_0}{\partial x} = 0 \tag{12}$$

$$p^0 : \quad \frac{\partial \alpha_0}{\partial t} - \alpha_1 \omega^2 x - \alpha_1 f(t)x^2 = 0 \tag{13}$$

The system (10) - (13) contains only three unknown functions α_0, α_1 and α_2 meaning it is overdetermined. Thus it becomes clear that integrability can not be expected for a general coefficient function $f(t)$. However we may obtain a formal solution of this system by solving Eq. (10) for α_2, using the so obtained result to resolve Eq. (11) and using α_2 and α_1 to extract α_0 from (12). Finally all these results are used in Eq. (13) to establish conditions on the coefficient function $f(t)$ in order to obtain a first integral.

Since the α_i are functions of x and t the general solution of the partial equation (10) is

$$\alpha_2 = a_2(t) \tag{14}$$

where $a_2(t)$ is a yet undefined function of time. Inserting this result into Eq. (11) and solving w.r.t. α_1 gives

$$\alpha_1(x,t) = -\dot{a}_2 x + a_3(t) \tag{15}$$

The function $\alpha_0(x,t)$ is found be substituting α_2 and α_1 into Eq. (12) and is

$$\alpha_0(x,t) = \frac{2}{3} a_2 f(t) x^3 + a_2 \omega^2 x^2 + \frac{1}{2} \ddot{a}_2 x^2 - \dot{a}_3 x + a_4(t) \tag{16}$$

Using these results in Eq. (13) and comparing equal powers in x we obtain a system of four ordinary differential equations in the four functions $a_2(t)$, $a_3(t)$, $a_4(t)$ and the coefficient function $f(t)$.

$$x^0: \quad \dot{a}_4 = 0 \tag{17}$$
$$x^1: \quad \ddot{a}_3 + \omega^2 a_3 = 0 \tag{18}$$
$$x^2: \quad \frac{d^3 a_2}{dt^3} + 4\dot{a}_2 \omega^2 - 2a_3 f(t) = 0 \tag{19}$$
$$x^3: \quad 5\dot{a}_2 f(t) + 2 a_2 \dot{f}(t) = 0 \tag{20}$$

The solution for a_4 is constant and contributes only to the constant of integration of the problem. The function a_3 is solution of the harmonic oscillator equation (18) and is given by

$$a_3(t) = C_1 \cos \omega t + C_2 \sin \omega t \tag{21}$$

where C_1 and C_2 are arbitrary real constants. For $a_2(t)$ we obtain a third order equation of the form

$$\frac{d^3 a_2}{dt^3} + 4\omega^2 \dot{a}_2 - (C_1 \cos \omega t + C_2 \sin \omega t) f(t) = 0 \tag{22}$$

while logarithmic integration of Eq. (20) leads to

$$5 \frac{\dot{a}_2}{a_2} = 2 \frac{\dot{f}}{f} \quad \Rightarrow \quad 5 \log a_2 = 2 \log f \quad \Rightarrow \quad f(t) = a_2(t)^{-5/2} \tag{23}$$

Using this condition for $f(t)$ we may rewrite (22) and we come to the following conclusion:

- The equation $\ddot{x} + \omega^2 x + f(t)x^2 = 0$ is integrable if $f(t)$ fulfills the condition

$$f(t) = a_2(t)^{-5/2} \tag{24}$$

where a_2 is solution of the generally nonlinear third order ordinary differential equation

$$\frac{d^3 a_2}{dt^3} + 4\omega^2 \dot{a}_2 - (C_1 \cos\omega t + C_2 \sin\omega t)a_2^{-5/2} = 0 \tag{25}$$

In this case the first integral reads as

$$I(x,p,t) = \frac{2}{3}a_2(t)^{-3/2}x^3 + \left[\frac{1}{2}\ddot{a}_2(t) + \omega^2 a_2(t)\right]x^2 - \dot{a}_3(t)x + [-\dot{a}_2(t)x + a_3(t)]p + a_2(t)p^2 \tag{26}$$

where

$$a_3(t) = C_1 \cos\omega t + C_2 \sin\omega t \quad ; \quad C_{1,2} \text{ are arbitrary real constants} \tag{27}$$

Following this statement we see that the function $f(t)$, although its form is prescribed, still depends on five parameters - $f(t) = f(t; K_1, K_2, K_3; C_1, C_2)$ -. The $K_{1,2,3}$ are the three integration constants resulting from solving the third order equation (25) while C_1 and C_2 enter this equation as free parameters.

A The special case $C_1 = C_2 = 0$

From the results just obtained it follows that finding a function $f(t)$ that renders Eq. (3) integrable we are oblidged to solve the third order nonlinear ordinary differential equation (25). However, it turns out that solving this equa tion might be as difficult as trying to do this for the original equation (3). On the other hand choosing the free constants C_1 and C_2 in Eq. (25) to be zero linearizes this equation and its solution becomes straightforward:

$$a_2(t)5 = a_2(t; C_1 = C_2 = 0) = K_3 + K_1 \cos 2\omega t + K_2 \sin 2\omega t \tag{28}$$

Then using Eq. (23) we find

- Equation $\ddot{x} + \omega^2 x + f(t)x^2 = 0$ is integrable if

$$f(t) = [K_3 + K_1 \cos 2\omega t + K_2 \sin 2\omega t]^{-5/2} \quad ; \quad K_3 > \sqrt{K_1^2 + K_2^2} \tag{29}$$

where the first integral is given by

$$I(x,p,t) = \frac{2}{3}a_2^{-3/2}(t)x^3 + \left[\frac{1}{2}\ddot{a}_2(t) + \omega^2 a_2(t)\right]x^2 - \dot{a}_2(t)xp + a_2(t)p^2 \tag{30}$$

and $a_2(t)$ is given by (28).

The geometrical interpretation of this integral is the one of an invariant surface with variable crossection in the extended phase space (x, p, t) on which the solution vector

$$\vec{X}(t) = \begin{pmatrix} x(t) \\ p(t) \end{pmatrix} \tag{31}$$

remains for any time t. The crossection repeats exactly after one period of the function $f(t)$, hence after a time equal to $T = \pi/\omega$. As the integral (30) is globally valid no chaotic motion can appear in this dynamical system [4].

IV NUMERICAL AND ANALYTICAL STUDY OF THE INTEGRABLE EQUATION

In order to illustrate the results obtained in the previous section, some numerically found soulutions of Eq. (3) are shown and discussed. In addition we derive an analytic expression for the ivariant curves in the (x, p) plane when $t = \pi/\omega$. Finally the stability limit in the frame of initial conditions is analytically established. First we show a graph of the function $f(t)$ when $K_3 = 1$, $K_1 = 0.3$, $K_2 = 0$ and $\omega = 0.7$.

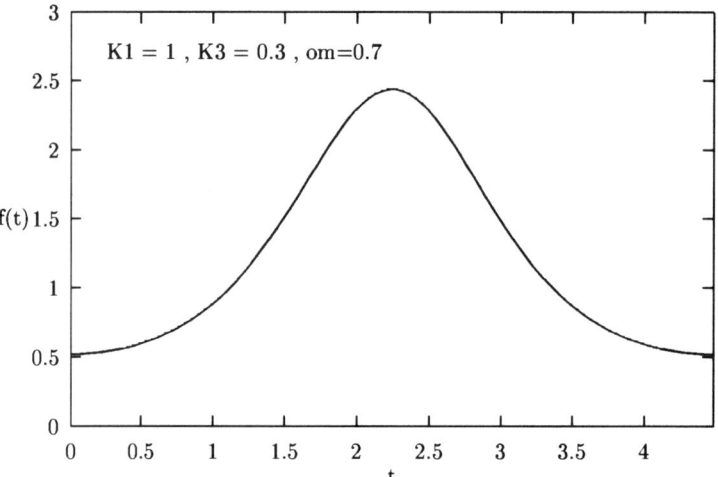

Fig. 1 Graph of the function $f(t)$ when $\omega = 0.7$, $K_3 = 1$, $K_1 = 0.3$ and $K_2 = 0$

For numeric investigations we use a fifth order Runge-Kutta integrator with automatic step size control [5] and show the so obtained numerical solutions for different initial conditions. The parameters $K_{1,2,3}$ are choosen to be $K_3 = 1$, $K_1 = 0.9$ and $K_2 = 0$.

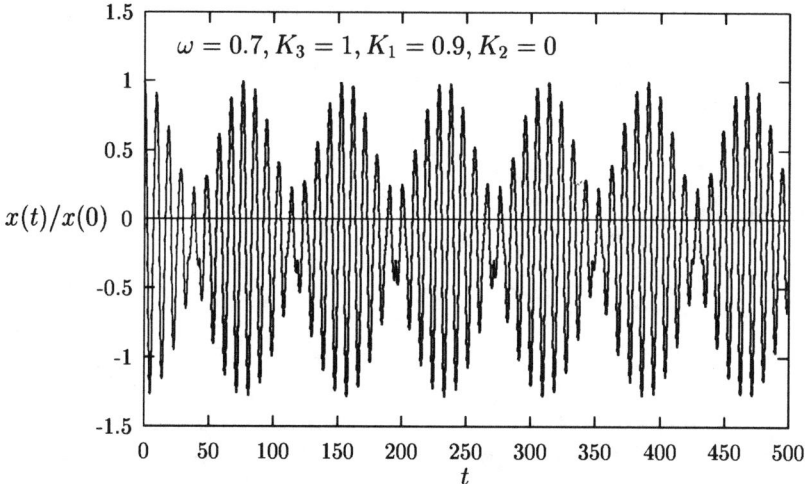

Fig. 2 Solution of equation for $x(0) = 0.04$, $p(0) = 0$

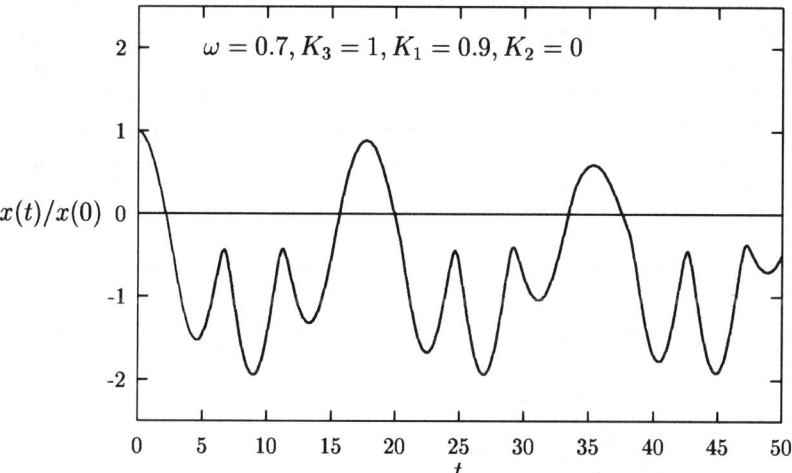

Fig. 3 Solution of equation for $x(0) = 0.0641$, $p(0) = 0$

The last two figures show the solution over a time interval of $0 < t < 500$ and $0 < t < 20$ for the initial conditions $x(0) = 0.04$ and $x(0) = 0.0641$. The initial momentum $p(0) = 0$ in both cases. In Fig.3 the 'maximum' bounded solution is shown, meaning that increasing the initial condition for x by only 10^{-4} leads to unbounded motion. In Fig.4 we show the first part of an unstable orbit (meaning unbounded in this case). We observe several oscillations before actually the solution starts to grow rapidly.

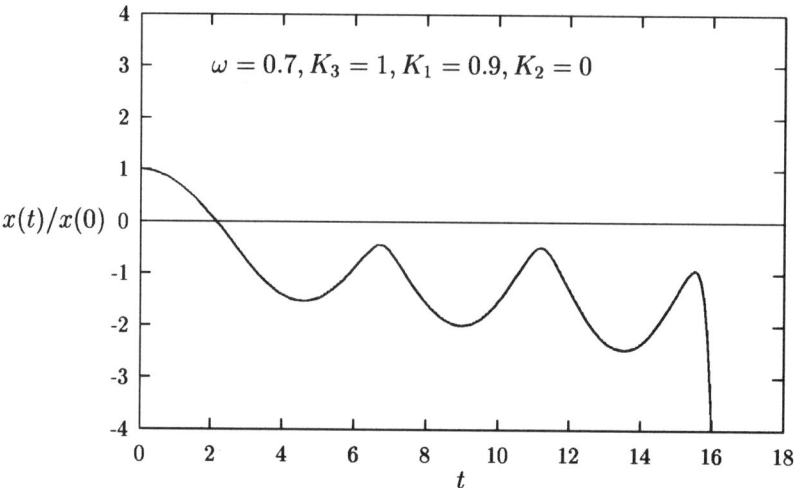

Fig. 4 Solution of equation for $x(0) = 0.0642$, $p(0) = 0$

While in the case of a relativly low amplitude (still far from the stability limit) the behaviour of the solution is essentially determined by two closeby frequencies leading to a strong beating, in the case close to the stability limit we see a solution strongly varying in amplitude and frequency meaning that many frequency components contribute to the solution. This situation is comparable to the behaviour of nonintegrable betatron motion close to the onset of chaotic and unbounded motion.

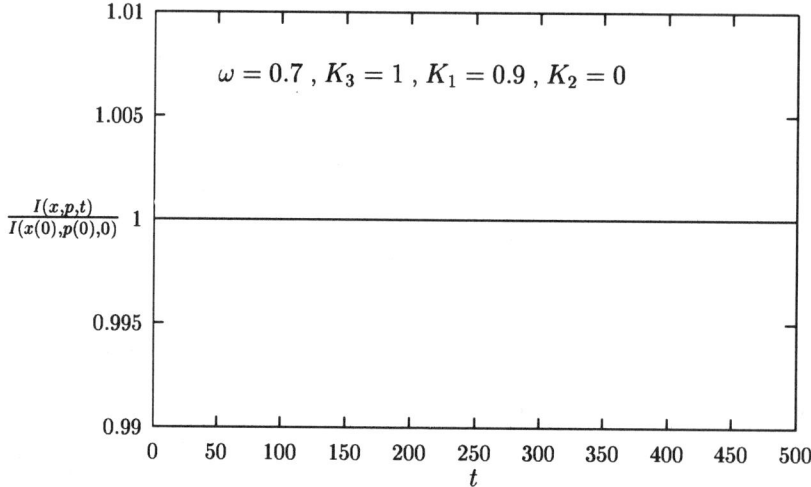

Fig. 5 Demonstration of the integral of motion for $x(0) = 0.0641$ and $p(0) = 0$

In Fig. 5 the actual solution $x(t)$ and its derivative $p(t)$ have been computed numerically for the last stable solution presented in Fig. 3 = and have been inserted into the first integral given by Eq. (30). We plotted $I(x,p,t)/I(x_0,p_0,0)$ an d over the entire interval of integration found a value exactly equal to unity.

Finally a Poincare surface of section plot of the extended phase space is shown. It can be seen that all the phase points $x(k\pi/\omega)$ and $p(k\pi/\omega)$ fall on well defined invariant curves shown in Fig. 6.

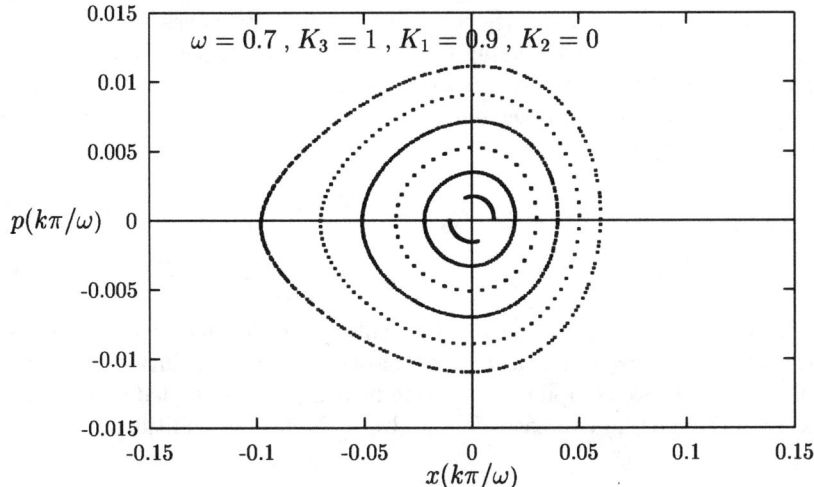

Fig. 6 Phase space surface of section plots for $x(0) = 0.01$ to 0.06

Evidently the analytic form of the invariants can be found by replacing $t = k\pi/\omega$ in (30). From (28) follows that (in case of $K_2 = 0$)

$$a_2(k\pi/\omega) = a_2(0) = K_3 + K_1 \tag{32}$$
$$\dot{a}_2(0) = 0 \tag{33}$$
$$\ddot{a}_2(0) = -4\omega^2 K_1 \tag{34}$$

Inserting these identities into (30) at $t = \pi/\omega$ results in

$$(K_3 - K_1)\omega^2 x^2 + \frac{2}{3}(K_3 + K_1)^{-3/2} x^3 + (K_3 + K_1)p^2 = C \tag{35}$$

where the constant C is equal to

$$C = (K_3 - K_1)\omega^2 x_0^2 + \frac{2}{3}(K_3 + K_1)^{-3/2} x_0^3 + (K_3 + K_1)p_0^2 \tag{36}$$

A Stability analysis

Given the analytic form of the invariant curves (35) it is straightforward to derive the maximum initial value for $x(0)$ and $p(0)$ that leads to a bounded motion. This problem is reduced to an algebraic procedure deciding wether the invariant in x and p rerlated to a certain set of initial conditions are closed curves or open to infinity. In addition, since the invariant curves repeat periodically in time and transform in a continous way as time increases, it is sufficient to study just one o f the curves e.g. the one for $t = k\pi/\omega$ given in (35). Here we specialize to the case $p(0) = 0$, hence we ask for the maximum initial value of x leading to bounded motion when the initial slope vanishes. Since the curves represented by (35) are symmetric w.r.t. the x axes (quadratic in p) we expect opening to infinity towards x direction. We therefore investigate the function

$$F(x) = \frac{2}{3}(K_3 + K_1)^{-3/2}x^3 + \omega^2(K_3 - K_1)x^2 - \frac{2}{3}(K_3 + K_1)^{-3/2}x_0^3 - \omega^2(K_3 - K_1)x_0^2 = 0 \quad (37)$$

with respect to the initial value x_0. This cubic equation has either three real solutions in which case the invariants are closed or one real solution (the initial condition x_0) and two complex solutions indicating an unbounded motion. The transition between the two cases follows from the usual conditions

$$F'(x_{Extr}) = 0 \Longrightarrow x_{Extr} \quad (38)$$
$$F(x_{Extr}) = 0 \quad (39)$$

which results in a cubic equation for the stability limit x_{lim}

$$\frac{2}{3}(K_3 + K_1)^{-3/2}x_{lim}^3 + \omega^2(K_3 - K_1)x_{lim}^2 - \frac{1}{3}\omega^6[(K_3 - K_1)(K_3 + K_1)]^3 = 0 \quad (40)$$

The actual solution to choose is the minimum of the real solutions which in all cases turns out to be

$$x_{lim} = \frac{1}{2}\omega^2(K_3 - K_1)(K_3 + K_1)^{3/2} \quad (41)$$

To check this formula we compare it to the results of numeric integration presented in Fig. 2 and 3. For $\omega = 0.7$, $K_3 = 1$ and $K_1 = 0.9$ we found that $x_0 = 0.0641$ leads to bounded motion up to $t = 500$ while $x_0 = 0.0642$ produced an unbounded solution. Inserting these parameters into Eq. (41). we find that indeed $x_{lim} = 0.06416....$ in full accordance with the numerical prediction.

V POSSIBLE APPLICATIONS TO A STORAGE RING

One of the major problems connected to the optical design of an accelerator or storage ring with nonlinear multipoles is the proper determination of the dynamical aperture. Although attempts have been made to determine the stability limit in analytic approaches (e.g. in [7]) a general theory has not yet been obtained and we still rely heavily on numerical tracking methods. It turns out to be a generic behaviour of nonlinear lattices to loose particles due to chaotic orbits which are possible because no global invariant curves covering the phase space can be found to which the particle motion is confined. Evidently particle loss is also possible in integrable cases (e.g the case of a single isolated resonance, [8]), but in these cases the mechanism is the one of an opening of invariant curves towards infinity, the particles being lost along these curves. In such a system the stability limit in amplitude can be obtained in closed form.

The basic idea of this chapter is to try to apply the nonlinear sextupolar distribution $f(t)$ found in the previous chapters (rendering the equation of motion integrable) to a real machine lattice. If we succeed to do this, not only it will be possible t o compute the dynamical aperture in an analytic way, but in addition NO fractional resonances will be excited. So the fractional part of the tune ω can in principle be choosen arbitrarely (e.g. equal to 1/3) without having to care about particle los ses. Following our example we therefore want to identify (see Eq. (1)

$$\frac{1}{2}\omega^2 K'(t)\beta(t)^{5/2} = [K_3 + K_1 \cos 2\omega t + K_2 \sin 2\omega t]^{-5/2} \qquad (42)$$

For simplicity we continue to use t as the independent variable and ω instead of the tune Q. Evidently there are three facts that limit us in applying exactly this function $f(t)$ to a real lattice:

- The general solution of a particle moving through sextupolar fields is two dimensional and Eq. (1) only is valid in the horizontal plane.

- Accelerator lattices are build up of lumped multipole distributions which by definition cannot be represented by a continous function.

- Any sextupole distribution in a circular accelerator has to close over the circumference, hence

$$f(t) = f(t + 2\pi) \qquad (43)$$

This is possible only if we choose ω to be

$$\omega = \frac{N}{2} \quad ; \quad N \cdots \text{integer} \qquad (44)$$

meaning we are located exactly on the linear stop band of the machine.

The first remark restricts the possible application of the proposed scheeme to a lattice for a flat beam (low vertical to horizontal emittance ratio).

Concerning the second point we are evidently obliged to use some approximation to the ideal sextupole distribution function. One possible way could be to use the three free parameters in $f(t)$ to adjust the chromaticity ω'_x to obtain a desired value as well as to minimize e.g. for the higher order chromaticities ω''_x and ω'''_x. Then one could cope for the non continous actual sextupole distribution by asking as many Fourier coefficients of the real distribution as possible to coincide with the ones of the continous function $f(t)$, hence

$$a_n = \frac{1}{\pi}\int_0^{2\pi} \frac{1}{2}\omega^2 K'(t)\beta(t)^{5/2} \cos nt\, dt = \frac{1}{\pi}\int_0^{2\pi} f(t)\cos nt\, dt \qquad (45)$$

$$b_n = \frac{1}{\pi}\int_0^{2\pi} \frac{1}{2}\omega^2 K'(t)\beta(t)^{5/2} \sin nt\, dt = \frac{1}{\pi}\int_0^{2\pi} f(t)\sin nt\, dt \qquad (46)$$

Of course asking for more coefficients to agree with the ones of the continous distribution will require more individually powered sextupoles along the lattice cells.

The third limitation mentioned seems to be the most critical one and we will investigate it in more detail. Since the periodicity 2π of $f(t)$ is obligatory, the only possibility in a practical design is to "stretch" the actual function, replacing

$$[\cos 2\omega t, \sin 2\omega t] \Longrightarrow [\cos 2(\omega + \epsilon)t, \sin 2(\omega + \epsilon)t] \qquad (47)$$

using for ϵ the smallest possible number to adchieve the required periodicity. For any given tune ω the minimum ϵ required is

$$\epsilon = Min_N \left[\frac{N}{2} - \omega\right] \qquad (48)$$

where N is one of the three integers $int(2\omega) - 1$, $int(2\omega)$ and $int(2\omega) + 1$. Since integrable systems are very isolated in the space of all possible dynamical systems we inevitably destroy the integrability of the equation of motion. However there is indication that choosing ϵ small, the integrability properties are only slightly perturbed. To show this we define the 'distance' D of a given dynamical system to some neighbouring integrable system by

$$D = [I(x,p,t;\epsilon), H(x,p,t;\epsilon)] + \frac{\partial I(x,p,t;\epsilon)}{\partial t} \qquad (49)$$

where $[I,H]$ is the usual Poisson bracket operator. The function $I(x,p,t;\epsilon)$ is obtained by performing the replacement (47) in the invariant of motion while $H(x,p,t;\epsilon)$ is the perturbed Hamiltonian. Evaluating this expression and expanding the result to first order in ϵ we find

$$D = -8\epsilon\omega^2 x^2[K_1 \sin 2\omega t + K_2 \cos 2\omega t] + O(\epsilon^2) \qquad (50)$$

Hence, as long as x is bounded, the deviation of the integrability properties of the dynamical system defined by the transformation (47) with respect to the integrable case is finite and decreases with ϵ. We therefore conclude that the dynamical aperture found in the "stretched" case (where $f(t)$ closes over 2π) will be close to the one of the integrable case and therefore can be well approximated by the formula given in (41). We check this result for the case $\omega = 7.33$ (note that we are located on a third integer resonance), $K_3 = 1$ and $K_1 = 0.5$ ($K_2 = 0$)In order to stretch the function $f(t)$ to have period 2π, $\epsilon = 0.17$ as follows from (48). The dynamical aperture for the case of $\epsilon = 0$ is given by $x_{lim} = 24.7$ as follows from (41). In order to obtain the stability limit for the perturbed case, we numerically integrate the perturbed equation

$$\ddot{x} + \omega^2 x + = [K_3 + K_1 \cos 2(\omega + \epsilon)t + K_2 \sin 2(\omega + \epsilon)t]^{-5/2} x^2 = 0 \qquad (51)$$

starting with increasing initial amplitudes (when $p(0) = 0$) untill the first occurrence of unbounded motion is observed. The integration interval has been extended over $1000 \times 2\pi$. The so obtained result is $x_{lim} = 23.7$ and thus differs by only 6% from the integrable case. The following two figures show a comparison between the phase space surface of section plots of the integrable case (Fig. 6) and the slightly nonintegrable case (Fig. 5). We clearly see that in the nonintegrable case invariant curves persist to exist up to the stability limit and chaotic motion is only appearing when the solution is already unstable.

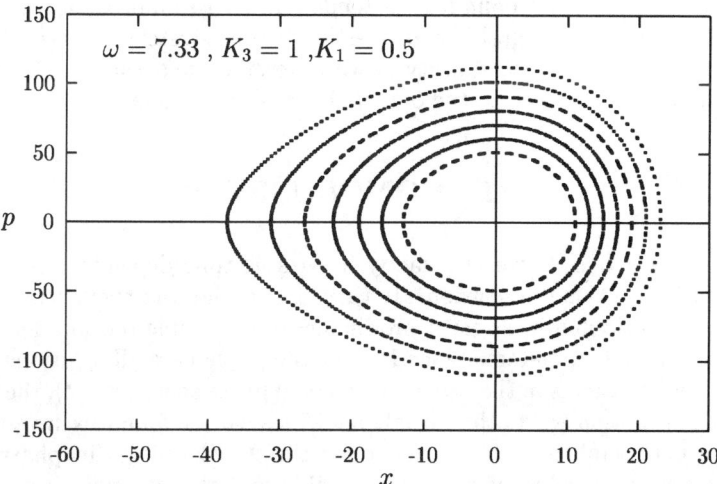

Fig.7 Phase space surface of section plot for the integrable case ($\epsilon = 0$)

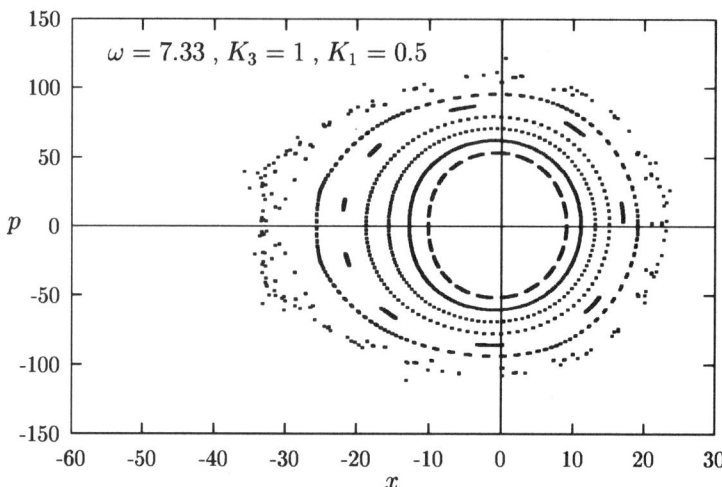

Fig.8 Phase space surface of section plot for the slightly non integrable case ($\epsilon = 0.17$)

Note that although the fractional part of the linear tune is close to 1/3, no resonance excitation to order three is visible in both cases. However, we observe a weak 7-th order resonance (islands) in the near integrable case when the initial amplitude is about 18. The presence of strong chaos **above** the stability limit is explained easily from the definition of the distance to the integrable system given in (50). Since this expression is proportional to x^2, once the solution is un bounded, D tends quickly to infinity thus destroying completly the near integrability properties of the "stretched" case. However, below the limit near integrability is well preserved and that is why the dynamical aperture is close to the one of the integrable equation.

VI CONCLUSIONS

We succeeded to derive an exactly integrable time dependent equation of betatron type in one dimension. It turned out that the coefficient function of the quadratic nonlinear term producing an integrable motion depends on three arbitrary parameters. Since in this integrable case all solutions remain on well defined curves in the three dimensional phase space (x, p, t), the limit of unbounded motion w.r.t. the initial conditions can be found by investigating up to which amplitude these curves remain closed objects in phase space. Following this procedure it has been possible to derive an analytic expression for this maximum amplitude.

Trying to apply this results to a real accelerator or storage ring we found that we have to modify slightly the coefficient function of the nonlinear term by changing its frequency and thus perturbing the integrability of the equation. However, we can show that for bounded solutions near integrability is preserved and thus the predicted dynamical aperture for the integrable case is close to the one found in the perturbed equation.

In addition the near integrable dynamical system is robust against strong isolated resonances (that act on low amplitudes) and this fact, together with the very simple procedure to find the dynamical aperture should encourage attempts to realize such a case in an accelerator or storage ring.

Acknowledgements: This research was supported in part by the *National Foundation under Grant No. PHY94-07194*, USA and by FEDER and Programa PRAXIS XXI, Projecto N° PRAXIS/2/2.1/FIS/176/94, Portugal.

REFERENCES

1. Courant, E.D., and Snyder, H.S. *Phys. Rev.* **88**, *1190*, (1952).
2. Hagel, J. and Bouquet,S., An integrable case of a time dependent second order equation with a quadratic nonlinearity, *Proc. NEEDS 1991, Baia Verde, Gallipoli.*
3. Goldstein, H. Classical Mechanics, *Addison-Wesley, Reading, Massachusetts, 1965.*
4. Lichtenberg, A.J. and Lieberman, M.A., Regular and stochastic motion, *Springer Verlag New York Heidelberg Berlin* , (1983).
5. Press,W. H., Flannery,B.P., Teukolsky, S.A. and Vetterling, W.T., Numerical recipes in FORTRAN, *Cambridge University Press*, (1989).
6. Private Communication.
7. Hagel, J. and Guignard, (1985)
8. Hagel, J. and Guignard, (1978)

Nonlinear Structures Near the Boundary of Marginal Stability

Tatiana A. Davydova, Alexei Yu. Pankin

Plasma Theory Department,
Institute for Nuclear Research,
pr. Nauki 47, Kiev - 22, 252022 Ukraine

Abstract. An explosive instability of the ion-temperature-gradient driven modes (η_i-modes) near the boundary of marginal stability is considered as driven mechanism of the subcritical turbulence. It is shown that a boundness of a wave interaction region leads to the saturation of the instability. Possibility of coherent soliton-like structures formation in both slab and toroidal geometry is demonstrated by numerical simulation. An analytical soliton solution is found in some special cases.

I INTRODUCTION

Formation of coherent long-lived large-scale structures of drift modes is sometimes considered to be one of the main reason of transition to the regime of improved confinement in tokamaks (L-H transition) [1] and sometimes as the reason for anomalous transport [2,3]. Though the connection between anomalous transport and coherent structures is still unclear, recent experiments indicate that "the correlation length which determines the local transport is determined by nonlinear effects" [5]. Recently new experimental evidences have been obtained that nonlinear wave-wave interaction between spontaneously excited quasi-coherent drift modes can lead to turbulence in a toroidal plasma [4]. The ion temperature gradient (ITG) driven modes along with the trapped electron mode and pressure gradient ballooning modes are dominating instabilities for the most realistic tokamak parameters (low β and weak collisionality) and in many cases are responsible for ion anomalous transport in tokamak plasmas [6]. The ITG driven mode (or η_i-mode) is driven by ion temperature gradients and is characterized by parameters $\eta_i = d\ln T/d\ln n$ and $\varepsilon_n = 2d\ln B/d\ln n$. A critical value of the parameter, η_{icr}, determines the threshold under which the mode is linearly stable [7]. Many experimental evidences [8,9] indicate that the tokamak plasma profiles are in line with the assumption that plasma is near the boundary of marginal stability for η_i-mode. Some kinetic simulations show also that the ITG driven mode is often close to linear stability boundary for measured tokamak parameters [10]. These results conform with profile consistency principle, that was put forward by Coppy [11], and supported later in [12,13]. One of the approaches to anomalous transport is that the increased transport will bring density $n(r)$ and temperature $T(r)$ profiles in plasma of tokamaks to marginal stability of strong reactive unstable modes such as the ion temperature gradient driven modes are. This could be considered as support of the idea of self-organization of tokamak plasmas. Below the linear stability boundary there still exists rather high level of turbulence, so called "subcritical turbulence". It is true not only for reactive drift modes, which ITG driven modes are. Subcritical turbulence has been observed in fluid dynamics [14] and in MHD plasmas [15].

Close to the stability boundary system behavior changes essentially and requests special analytical treatment. The nonlinear explosive instability due to the interaction between modes with positive and negative energy have proposed as the driving mechanism for the subcritical turbulence [16]. As shown in [17] the character of nonlinear wave interaction radically changes near the stability boundary. When all interacting modes are "zero energy modes" characteristic nonlinear interaction time t_0 is the smallest [18]. By "zero energy wave" we mean a quasi-monochromatic wave of frequency ω_k for which the following conditions hold:

$$\omega_k^{-1} \ll \tau_{nl} \ll \left| \frac{1}{E(\omega_k)} \left[\frac{dE(\omega)}{d\omega} \right]_{\omega=\omega_k} \right|, \tag{1}$$

where τ_{nl} is the characteristic time of nonlinear process and $E(\omega)$ is the energy of the quasi-monochromatic wave $E(\omega_k) = A^2 \omega_k [\partial D/\partial \omega]_{\omega=\omega_k}$, $D(\omega,k) = 0$ is the linear dispersion equation, and A is the wave amplitude. Inequalities (1) are consistent for a sufficiently large wave amplitude near the stability boundary of a wave with frequency ω_k $\left(D(\omega_k,k) = 0, [\partial D/\partial \omega]_{\omega=\omega_k} = 0\right)$. Then mode energy exchange during nonlinear interaction greatly exceeds "own" energies of modes. Other important feature of nonlinear wave interaction near the stability boundary is that the possibility of explosive instability does not depend on the signs of "own" wave energies as well as on the signs of interaction matrix elements if two or three waves from resonant triad are "zero energy waves" [19].

The purpose of the present paper is to investigate nonlinear explosive instabilities of η_i-modes near the boundary of marginal stability and their possible saturation mechanisms. It is known a number of saturation mechanisms of explosive instabilities. If free energy is kept unaltered at the expense of the energy of external sources, then self-stabilization of the wave-wave interaction by a nonlinear frequency shift (in the second order of perturbation theory) is possible [22]. The higher order nonlinear effects restrict the increase of interacting wave amplitudes at rather high amplitudes. The boundedness of the wave interaction region due either to the inhomogeneity of the medium, leading to detuning of the wave phases, or to the boundedness of the system may affect the explosive instability dynamic. In this case the explosive instability is saturated at sufficiently lower wave amplitudes if unstable perturbations escape from the interaction region in a smaller time comparing to the "explosion time". The stabilization conditions for the usual explosive instability due to medium inhomogeneity have been found in [20]. Below some intensity wave threshold only finite spatial wave amplification takes place. This conclusion was applied in [21] to explain formation of small-scale structures in the auroral plasma. In the case of modified explosive instability (1) frequency mismatch in inhomogeneous media gives a threshold to the instabilities only if the signs of matrix elements are equal [19]. As for different signs of matrix elements, temporal "explosion" of "zero-energy waves" does not stop in the first order of perturbation theory [19]. We show here that account for finite size of interaction area is more restrictive; that is that the instability threshold exists for any signs of the matrix elements.

The paper is organized as follows. In section 2 we outline a hydrodynamic slab model of η_i-modes near the boundary of marginal stability. In section 3 we review briefly time development of modified explosive instability of three resonantly coupling η_i-modes and in section 4 we show how the development is modified accounting for a finite size of interaction area. We demonstrate analytically and numerically that in bounded region L evolution of explosive instability of ITG driven modes near the boundary of marginal stability may lead to the steady state if $\tilde{L}$ is less than the "explosive" length for any signs of matrix elements. Finite interaction area of interacting wave packets is created in the self-organized process during spatial temporal instability evolution. This process may lead to drift wave envelope nonlinear structures (solitons) formation. In section 5 we find particular analytical solutions in the form of solitons and confirm numerically an existence of soliton solutions in cases of homogeneous as well as inhomogeneous background parameters. Section 6 contains discussion of our results.

II BASIC EQUATIONS

A wide range of theoretical descriptions of η_i-mode has been made in the last decades. The main treatments are: local [23,24] and nonlocal [25,26] approximations, fluid [7,27] and kinetic [24,28] description, in slab [7,28] and toroidal [25,26] geometry. We explore here a reactive slab fluid model in electrostatic approximation [7]. The basic set of equations consists of the ion continuity equation, and the ion momentum and energy equations. Compared to earlier fluid models [27], the present model combines simplicity with satisfactory description which is in an agreement with the results of kinetic model. The model takes into account all inhomogeneities (density, temperature and magnetic field) in the radial ($\hat{x}$)-direction, finite Larmore radius effects and difference between ion and electron temperatures. We put for simplicity here $\tau = T_e/T_i = 1$. Then in a Fourier-representation the model equations take the form:

$$\left(i\frac{\partial}{\partial t} + k_y A\right) T_k + k_y B \Phi_k = i \sum_{k_1,k_2} \left(V^{(1)}_{kk_1k_2} \Phi^*_{k_1} \Phi^*_{k_2} + W^{(1)}_{kk_1k_2} T^*_{k_1} \Phi^*_{k_2}\right)$$

$$\left(i\frac{\partial}{\partial t} + k_y C\right) \Phi_k + k_y D T_k = i \sum_{k_1,k_2} \left(V^{(2)}_{kk_1k_2} \Phi^*_{k_1} \Phi^*_{k_2} + W^{(2)}_{kk_1k_2} T^*_{k_1} \Phi^*_{k_2}\right) \quad (2)$$

where T_{k_j} and Φ_{k_j} are Fourier components of ion temperature and electrostatic potential perturbations which are normalized as follows

$$\hat{T} = \frac{L_n}{\rho_s} \frac{\delta T}{T}, \quad \hat{\Phi} = \frac{L_n}{\rho_s} \frac{e\Phi}{T}, \quad T = T_e = T_i.$$

Here we have used the notations:

$$A = \frac{\varepsilon_n}{3}\left(7 - 2k^2\right),$$

$$B = \frac{4\varepsilon_n}{3}\left(1 - k^2\right) - \eta_i + \frac{2}{3}(2 + \eta_i)k^2,$$

$$C = \varepsilon_n \left(1 - k^2\right),$$

$$D = 2C - \frac{1 - (1 + \eta_i) k^2}{1 + k^2},$$

$$V^{(1)}_{kk_1k_2} = \frac{i}{3}\left(\vec{k_1} \times \vec{k_2}\right) \cdot \vec{e_\parallel} \left(k_2^2 - k_1^2\right),$$

$$W^{(1)}_{kk_1k_2} = \frac{i}{3}\left(\vec{k_1} \times \vec{k_2}\right) \cdot \vec{e_\parallel} \left(k_2^2 - \frac{3}{2}\right),$$

$$V^{(2)}_{kk_1k_2} = \frac{i}{2}\left(\vec{k_1} \times \vec{k_2}\right) \cdot \vec{e_\parallel} \frac{k_2^2 - k_1^2}{1 + k^2},$$

$$W^{(2)}_{kk_1k_2} = \frac{i}{2}\left(\vec{k_1} \times \vec{k_2}\right) \cdot \vec{e_\parallel} \frac{k_2^2}{1 + k^2},$$

$$\omega_s = c_s/L_n, \quad \rho_s = c_s/\Omega_{ci}, \quad L_n = \frac{d \ln n}{dx}.$$

In the linear approximation one can put $\Phi_{k_j}, T_{k_j} \sim \exp\left[-i\omega_{k_j} t\right]$ and obtain the linear dispersion relation for η_i-modes

$$\omega^{\pm}_{k_j} = \frac{1}{2}k_{yj} (A + D) \pm \frac{1}{2}k_{yj} \left((A - D)^2 + 4BC\right)^{1/2}$$

$$= \frac{1}{2}k_{yj}\left(1 - \frac{13}{3}\varepsilon_n\right) - \frac{3}{8}k_{yj}^2 \left(\varepsilon_n + \eta_i\right) \pm \frac{1}{2}k_{yj}\delta_k \quad (3)$$

where $\delta_j = 2\left[\varepsilon_n(\eta_{icr} - \eta_i)\right]^{1/2}$,

$\eta_{icr} = \eta_{icr}^0 + k^2\left(2 - \frac{5}{18}\varepsilon_n - \frac{1}{2\varepsilon_n} - \frac{\eta_{icr}^0}{2}\left(1 + \frac{1}{\varepsilon_n}\right)\right)$, $\eta_{icr}^0 = \frac{1}{6} + \frac{49}{36}\varepsilon_n + \frac{1}{4\varepsilon_n}$.

For weak nonlinearities ($\tau_{nl} \gg \omega_k^{-1}$) the basic set of equations for amplitudes of three resonantly interacting waves for which

$$\omega_{k_1} + \omega_{k_2} + \omega_{k_3} = 0, \ \vec{k_1} + \vec{k_2} + \vec{k_3} = 0, \quad (4)$$

reduces to the set of three equations:

$$\begin{aligned}\hat{L}_1 \Phi_2 &= V_{k_1 k_2 k_3} \Phi_2^* \Phi_3^* \\ \hat{L}_2 \Phi_3 &= V_{k_2 k_3 k_1} \Phi_1^* \Phi_3^* \\ \hat{L}_3 \Phi_3 &= V_{k_3 k_1 k_2} \Phi_1^* \Phi_2^* \end{aligned} \quad (5)$$

where

$$\hat{L}_j = \pm ik_{yj}\delta_j\left(\frac{\partial}{\partial t} - u_j\frac{\partial}{\partial y}\right) + \left(\frac{\partial}{\partial t} - u_j\frac{\partial}{\partial y}\right)^2$$
$$- 2ia_j\left(k_{xj}\frac{\partial}{\partial x} + k_{yj}\frac{\partial}{\partial y}\right) - a_j\left(\frac{\partial^2}{\partial x^2} + \frac{\partial^2}{\partial y^2}\right) \quad (6)$$

and

$$V_{k_i k_j k_l} = ik_{yj}\left(\vec{k_j} \times \vec{k_i}\right) \cdot \vec{e_\parallel}\left(k_j^2 - k_i^2\right)V,$$
$$a_j = k_{yj}^2\left(-u_j\left(\eta_i + \frac{16}{3}\varepsilon_n\right) + 10\varepsilon_n + 8\eta_i - 2\right),$$
$$V = 2\varepsilon_n - \frac{2}{3} - \frac{\eta_i}{4}, \ u_j = \frac{\omega_j}{k_{yj}}.$$

Signs $\pm$ before the first term in (6) correspond to two branches of the linear η_i-modes (3).

Sufficiently far from the boundary of marginal stability or for sufficiently small wave amplitudes, when inequality opposite to (1) is satisfied, the system (5) describes conventional explosive instability. For our problem this is valid if

$$\tau_{nl}\left(\omega_k^+ - \omega_k^-\right) \gg 1 \text{ or } \tau_{nl}\varepsilon_n^{1/2}(\eta_{icr}^0 - \eta_i) \gg 1.$$

The temporal development of this instability for η_i-modes was considered before in [16]. In the present paper we consider explosive instability in the other limiting case:

$$\tau_{nl}(\omega_k^+ - \omega_k^-) \ll 1 \text{ or } \tau_{nl}\varepsilon_n^{1/2}(\eta_{icr}^0 - \eta_i) \ll 1$$

when inequalities (1) hold.

III TEMPORAL DEVELOPMENT OF EXPLOSIVE INSTABILITY NEAR THE BOUNDARY OF MARGINAL STABILITY

Near the boundary of marginal stability the temporal evolution of nonlinear instability is described by

$$\frac{d^2\chi_j}{dt^2} = W_{k_j k_l k_m} \chi_l^* \chi_m^*, \qquad (7)$$

where $\chi_j = i\Phi_{k_j}$, and matrix elements $W_{k_j k_l k_m} = iV_{k_j k_l k_m}$ are real. Indexes j, l and m are running through the values (1,2,3), (2,3,1) and (3,1,2) respectively.

The system has self-similar "explosive" solutions for equal signs of matrix elements [18,19]:

$$\chi_i = \frac{6}{\sqrt{|W_{k_j k_i k_l} W_{k_i k_j k_l}|}(t_0 - t)^2}, \qquad (8)$$

and for different signs of matrix elements $(sign(W_{k_1 k_2 k_3}) = -sign(W_{k_2 k_3 k_1}) = -sign(W_{k_3 k_2 k_1}))$ [19]:

$$\chi_1 = \frac{\sqrt{84}\exp(i\alpha)}{\sqrt{|W_{k_2 k_3 k_1} W_{k_3 k_2 k_1}|}(t_0 - t)^{2(1-i\sqrt{3})}},$$

$$\chi_2 = \chi_3 \sqrt{\left|\frac{W_{k_2 k_3 k_1}}{W_{k_3 k_2 k_1}}\right|} = \frac{\sqrt{168}\exp(i\beta)}{\sqrt{|W_{k_3 k_2 k_1} W_{k_1 k_2 k_3}|}(t_0 - t)^{2-i\sqrt{3}}} \qquad (9)$$

where α and β are constants.

Except for the "explosive" solutions the system (7), with different signs of matrix elements, has also solutions (Fig. 1a) with one growing mode fitted approximately to $\chi_1 \sim at^{4/3}$ and two decaying oscillating modes fitted to $\chi_2, \chi_3 \sim \sqrt{8a/3}\, t^{-1/3} \cos\left(\frac{3}{5}\sqrt{a}t^{5/3} + \theta_0\right)$, where a, θ_0 are constants. In general case of the system (7) with solution (Fig. 1b) consist in a combination of "explosive" and "non-explosive".

The arbitrary solution of system (7) with equal signs of matrix elements tends to self-similar solution (8). Solutions (8) and (9) describe modified explosive instability when all three interacting waves are "zero energy waves". The amplitudes of all modes during an explosive instability simultaneously increase due to the sources of the medium nonequilibrium. In our case it is an inhomogeneity of background plasma parameters.

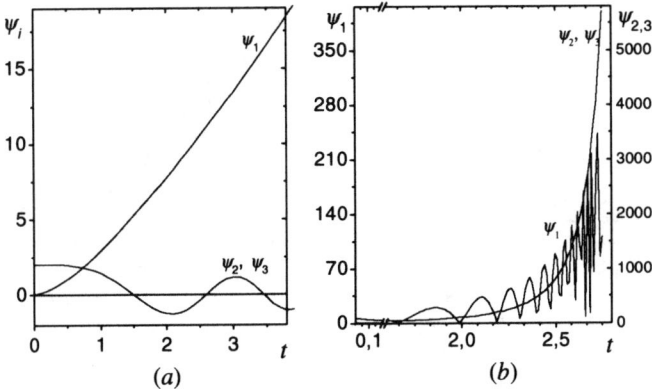

FIGURE 1. Typical numerical solutions $\chi_j = \psi_j \exp(i\varphi_j)$ of the system (7).

IV MODIFIED EXPLOSIVE INSTABILITY SATURATION IN A BOUNDED DOMAIN

Temporal description of instability is valid only if one can consider interacting waves with amplitudes, which does not depend on space coordinates. In a reality this is not the case and we should solve spatial-temporal problem to describe wave amplitudes' evolution. In this case an obvious generalization of the system (7) which follows from (5):

$$\left(\frac{\partial}{\partial t} - u_j \frac{\partial}{\partial x}\right)^2 \chi_j = V^*_{k_j k_l k_m} \chi^*_l \chi^*_m \tag{10}$$

At first let us consider simplified case of equal and real wave amplitudes $\chi_j = \chi$, equal phase velocities $u_j = u$ and equal signs of matrix elements. Then the system (10) reduces to the equation:

$$\left(\frac{\partial}{\partial t} - u\frac{\partial}{\partial x}\right)^2 \chi = \chi^2 \tag{11}$$

Here the amplitudes are renormalized so that $V_{k_j k_l k_m} = 1$. Introducing new function $\chi^{-1/2}$ and then using Laplace time transform it can be shown, that the equation (11) has the following solution in the interval $0 \leq x \leq L$:

$$\chi(x,t) = \begin{cases} \frac{6}{(t_0-t)^2}, & 0 \leq t < \frac{x}{u}, \; t < \sqrt{\frac{6}{a}} = t_0 \\ \frac{6u^2}{(l_0-x)^2}, & t > \frac{x}{u}, \; x < \sqrt{\frac{6}{b}}u = l_0 \end{cases} \tag{12}$$

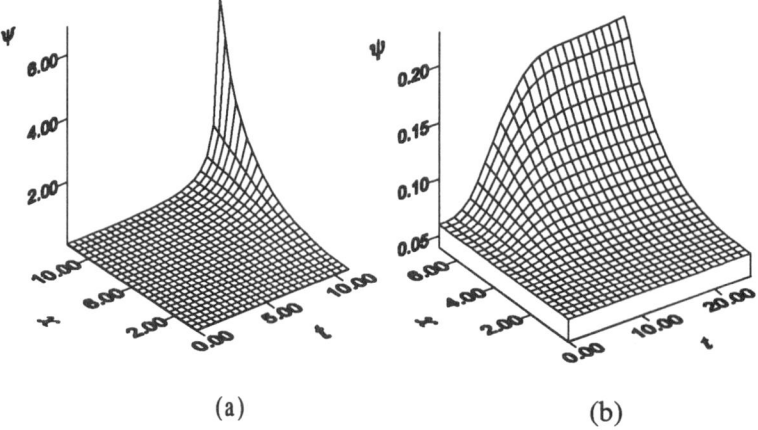

FIGURE 2. Two alternatives for system (10) with $t_{expl} \approx 13.5$, $L_{expl} \approx 12.8$ subject to the relation of system length L and "explosive" length L_{expl}: (a) Explosion ($L > L_{expl}$); (b) Saturation ($L < L_{expl}$).

for the following initial and boundary conditions: $\chi(x,0) = a$, $\chi(0,t) = b$, with $a < b$. Hence for $t > L/u$ the steady state on interval L is established in the nonlinear system (11) if $L < l_0 < t_0 u$, where l_0 is an "explosive length" of the instability for the corresponding "spatial" problem of the system (10). The steady state describes amplitude spatial distribution corresponding to the finite spatial amplification. We have found numerically for other boundary and initial conditions that during the temporal instability evolution different initial amplitudes tend to become equal and the phase difference $(\varphi_1 - \varphi_2 - \varphi_3)$ tends to zero as $t \to t_0$. Numerical calculations confirm that for given boundary conditions at the $x = 0$ on the finite interval $(0, L)$ steady state is established in the time $L/\min(u_j)$ if L is less than the "explosive length" l_0. We have checked this conclusion numerically for a wide spectrum of initial and boundary conditions and for any signs of matrix elements $V_{k_j k_l k_m}$ as well as for different signs of u_j (see an example on Fig. 2). Fig. 2a demonstrates explosive instability in the case $L > l_0$. Fig. 2b corresponds to the case $L < l_0$ when the steady state is established after the time $L/\min(u_j)$.

The model (10) may be applied to nonlinear wave interactions near boundary of marginal stability in various non-equilibrium systems. One of such systems is a plasma with monoenergetic ion beam [29]. This system becomes linearly unstable if the velocity of ion beam u is less than the sound velocity c_s or more strictly if $u < c_s (1 + \eta)^3$, where $\eta = n_b/n \ll 1$, n, n_b are densities of plasma and ion beam respectively. If $u \lesssim c_s (1 + \eta)^3$ the system is near the boundary of marginal stability. In the experiment [30] two probe waves were injected in a plasma-ion beam system and the third eigen mode was excited so that the matching conditions (4) were fulfilled. The simultaneous amplitude growth of all three waves was detected which is characteristic for explosive instability. This instability was interpreted in [18] as the explosive instability of "zero energy waves". It explains adequately observed characteristic time scales of temporal waves evolution.

Though our consideration above is evidently too simplified for η_i-modes in tokamaks but it demonstrates that inhomogeneity may play a role of a stabilizing factor leading to a steady state formation.

V ENVELOPE SOLITON SOLUTIONS

A Case of homogeneous background parameters

In this subsection we seek for steady solutions of the full system (5). We consider the possibility for localized solutions (envelope solitons) moving with constant velocity U along the direction of drift wave propagation ($\hat{y}$). To do this we introduce new variable $\xi = y + Ut$ and assume that the envelope soliton has the form:

$$\chi_j = \psi_j(\xi) e^{i(s_j y + q_j x)}, \quad (13)$$

with parameters $s_j \ll k_{yj}$ and $q_j \ll k_{xj}$, which satisfy the conditions:

$$s_1 + s_2 + s_3 = 0, \quad (14)$$
$$q_1 + q_2 + q_3 = 0. \quad (15)$$

Then the real functions ψ_j satisfy the following set of ordinary differential equations:

$$\left(1 + \frac{(U - u_j)^2}{a_j}\right) \frac{d^2 \psi_j}{d\xi^2} - 2i \left(\frac{u_j (U - u_j) s_j}{a_j} + \frac{k_{yj} \delta_j (U - u_j)}{2 a_j} + k_{yj} - s_j\right) \frac{d\psi_j}{d\xi}$$
$$- \left(\frac{u_j^2 s_j^2}{a_j} - 2k_{xj} q_j - 2k_{yj} s_j\right) \psi_j = \frac{V_{k_j k_l k_m}}{a_j} \psi_l \psi_m \quad (16)$$

On putting the second coefficient in the system (16) equal to zero, we arrive at the set of equations:

$$s_j \left(1 - \frac{u_j (U - u_j)}{a_j}\right) = k_{yj} \left(1 + \frac{1}{2} \frac{\delta_j}{a_j} (U - u_j)\right) \cong k_{yj}, \quad (17)$$

which along with the condition (14) determines two possible soliton velocities $U^{\pm}$:

$$U^{\pm} \approx \frac{k_{y1}^3 (V_2 + V_3) + k_{y2}^3 (V_1 + V_3) + k_{y3}^3 (V_2 + V_1) \pm \sqrt{D}}{2 \left(k_{y1}^3 + k_{y2}^3 + k_{y3}^3\right)}, \quad (18)$$

where $V_j \approx u_j + \frac{a_j}{u_j}$ and

$$D = \left(k_{y1}^3 (V_2 - V_3) + k_{y2}^3 (V_3 - V_1) + k_{y3}^3 (V_1 - V_2)\right)^2$$
$$+ 4k_{y2}^3 k_{y3}^3 (V_1 - V_2)(V_1 - V_3).$$

Then from (17) one could find parameters s_j.

One can see, that $D > 0$ and $U^{\pm}$ are real in all instances. Soliton velocity $U^{\pm}$ are of the order of phase velocity of drift waves. Next, we denote

$$\lambda_j = \frac{\left(2k_{xj}q_j + 2k_{yj}s_j - \frac{u_j^2 s_j^2}{a_j}\right)}{\left(1 + \frac{(U - u_j)^2}{a_j}\right)}, \quad (19)$$

As a result, the system (16) after the renormalization:

$$\psi_j = -\frac{\phi_j \sqrt{a_l a_m}}{\sqrt{\left|W_{k_l k_m k_j} W_{k_m k_j k_l}\right|}}$$

takes a simple form:

$$\frac{d^2 \phi_j}{d\xi^2} + \lambda_j \phi_j + S_j \phi_l \phi_m = 0, \quad (20)$$

where $S_j = sign\left(W_{k_j k_l k_m}\right)$. So far soliton parameters q_j are still arbitrary.

There are two principally different cases depending on signs of matrix elements. If all signs are equal ($S_1 = S_2 = S_3 = 1$) and all parameters λ_j are equal

$$\lambda_1 = \lambda_2 = \lambda_3 = -\lambda^2 \quad (21)$$

the system (20) has the particular analytical solution in the form of KdV soliton:

$$\phi_1 = \phi_2 = \phi_3 = \frac{3\lambda^2}{2 \cosh^2 \left(\lambda (y + Ut - y_0)/2\right)}. \quad (22)$$

Otherwise, when the signs of matrix elements are different ($-S_1 = S_2 = S_3 = 1$) and parameters λ_j relate as

$$-\lambda_1 = \lambda_2/2 = \lambda_3/2 = -\lambda^2 \quad (23)$$

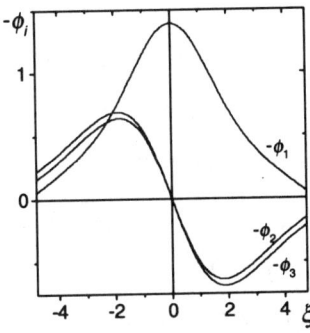

FIGURE 3. Soliton solution of system (20) with $\lambda_1 = -0.4$, $\lambda_2 = 0.25$, $\lambda_3 = 0.35$.

the system (20) has the soliton solution in the form:

$$\phi_1 = 6\lambda^2 \frac{\sinh(\lambda(y + Ut - y_0))}{\cosh^2(\lambda(y + Ut - y_0))},$$

$$\phi_{2,3} = -\frac{3\lambda^2}{\cosh^2(\lambda(y + Ut - y_0))}. \qquad (24)$$

Soliton parameters q_j and λ in both cases (22) and (24) are determined from the equation (15) complemented by three additional conditions for λ_j (21) or (23). For the solution (24) soliton intensity $|\phi_1|^2$ of the first mode has two bumps contrary to the two other drift soliton envelopes $|\phi_2|^2$ and $|\phi_3|^2$. Soliton solutions of the system (20) with arbitrary values of q_j (and hence λ_j) can be found numerically. In Fig. 3 an example of the numerical solution of the system (20) with unequal values of λ_j is presented. This case is similar to the case (23), considered analytically. The numerical simulation confirms that soliton solutions with two bumps are typical for the system (20) with different signs of λ_j.

Thus for fixed coefficients of the system (16) envelope drift solitons may propagate with two different velocities (18). The similar property is valid for small amplitude linear η_i-modes: two eigen modes (3) with the same wave vector propagate with different phase and group velocities. In the process of nonlinear interaction they are "merging" into "zero energy wave" but double nature of latter wave manifests itself after the self-saturation. This feature is likely to be inherent to nonlinear evolution of "zero-energy" or reactively unstable modes. For example, two nonlinear localized wave structures are formed on nonlinear self-stabilizing stage of electron beam-plasma instability. Such structures have been experimentally observed in [31] to move with slightly different velocities of order of the group velocity v_g of the most unstable linear mode ($v_g = 2v_0/3$, where v_0 is the electron beam velocity). This experiment was theoretically interpreted in [32].

B Case of inhomogeneous background parameters

Up to now we supposed that all background parameters of our model nonlinear system (all coefficients) are homogeneous. In general case two-dimensional inhomogeneity of magnetic tokamak field is of great importance for the formation of structure elements of turbulence. Linear localized ITG structures, so called ballooning ITG modes, are formed in a tokamak magnetic field [25,26]. Short-scale structures are usually described in strong ballooning approximation. Width of these structures in poloidal direction is determined by Schrödinger equation eigen problem with square potential well $(U(\theta) = U_0(\theta_0) - \kappa(\theta_0)(\theta - \theta_0)^2$, where θ_0 is an extremum of effective potential function). The most deep potential well is usually formed at the outer part of the discharge ($\theta_0 = 0$, and $\kappa(\theta_0) = 0$), leading to the ballooning nature of linear global drift modes. Experiments and simulations point out that characteristic size of structures can not be explained only by linear effects [5]. As we have shown in the previous subsection nonlinear coupling may lead to localized structures formation even in the case of homogeneous background parameters. The influence both linear effects, or an existence of effective potential well determined by tokamak magnetic field geometry, and nonlinear wave-wave interaction effects on ITG structures is of great interest. In particular, we try to elucidate whether localized drift structures may be formed on the inner side of the torus ($\theta_0 = \pi$, $\kappa(\theta_0) < 0$) due to nonlinear coupling. In this subsection we use the nonlinear system (20) taking into account dependence of parameters λ_j on poloidal ($\widehat{y}$) direction near the extremum points ($\lambda_j \to \lambda_j + \kappa_j y^2$):

$$\frac{d^2\chi_j}{d\xi^2} + \left[\lambda_j + \kappa_j \xi^2\right]\chi_j + S_j \chi_l \chi_m = 0, \tag{25}$$

where χ_j is connected with the drift wave potentials Φ_{k_j} by $\chi_j = i\Phi_{k_j}$ and $\xi = y + Ut$. Here we assume that envelopes of interacting drift modes potentials χ_j are localized near the point $y = 0$ where the parabolic approximation of λ_j dependence on y is valid.

In linear approximation three equations of the system (25) are independent. Simple analysis [25] shows that in this case well-localized mode in $\widehat{y}$-direction may exist for negative κ_j and positive discrete eigenvalues λ_j. In the case of positive κ_j the corresponding linear problem (25) was interpreted in [25] as describing extended modes escaping out of the point $y = 0$ in both directions with imaginary eigenvalues λ_j. However nonlinear wave coupling can essentially change this picture. We have shown that the system (25) admits

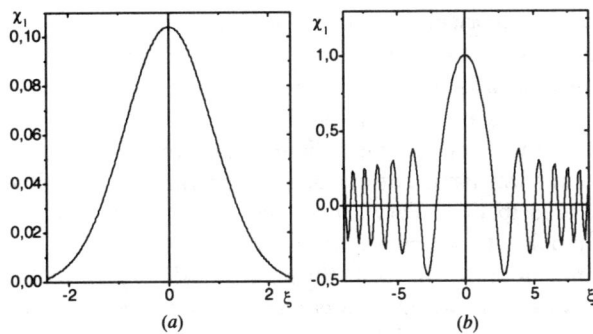

FIGURE 4. Amplitude of first η_i-mode as solution of the system (25) with: (a) $\lambda_{1,2} = -1.26$, $\lambda_3 = -1.17$, $\kappa_1 = -1.7$, $\kappa_2 = -2.3$, $\kappa_3 = -1.44$, $V_j = 1$; (b) $\lambda_1 = 1.2$, $\lambda_2 = 1$, $\lambda_3 = 0.8$, $\kappa_1 = 0.9$, $\kappa_2 = 5$, $\kappa_3 = 1.1$, $V_j = 1$.

stationary localized solutions both for $\kappa_j > 0$ and $\kappa_j < 0$. Results of the numerical simulations (Fig. 4) reveal that two possible types of nonlinear structures can appear in such system. Contrary to the case of linearized system monotonically decaying localized solution (Fig. 4a) have been found for parameters $\kappa_j < 0$ and $\lambda_j < 0$. Envelope soliton solution with oscillating tails (Fig. 4b) corresponds the case $(\kappa_j > 0)$, when the linearized system (25) have no localized solution too.

VI DISCUSSION

We have shown that near the boundary of marginal stability nonlinear explosive instability of ITG driven modes develops even if the plasma is linearly stable. The instability may result in coherent η_i-structures, that constitute the structure of subcritical turbulence. An explanation of the subcritical plasma turbulence and anomalous transport was advanced by Itoh et al. [33,34] on the MHD model of electrostatic current diffusive intercharge mode. In their "self-sustained turbulence" theory the nonlinear marginal stability condition determines the anomalous transport coefficient. They stressed the nonlinear instability as a governing factor for anomalous transport. However contrary to our approach they supposed that excited modes are completely decorrelated and uniformly distributed in space. Then the self-stabilization of the instability occurs because of enhanced transport. We believe turbulent state arise after the saturation of the instability and appearance of the coherent structures. The turbulent state is formed in interactions (collisions) of these structures. Some numerical 2D simulations confirm this point of view [1].

In the framework of a simple three-wave interaction model we have shown that formation of spatially nonuniform structures is a plausible saturation mechanism for nonlinear instabilities. We have used a one-dimensional nonlinear model. It may happen that these structures turn out to be unstable against perpendicular perturbations as it is the case for one dimensional drift solitons discovered in [35]. Petviashvili has shown that this instability leads to stable monopolar two-dimensional drift solitons or vortices formation in the case of scalar nonlinearity [36]. Such vortices has been observed in model experiment on water in rotating vessel and in Earth's magnetosphere [37]. We expect that two-dimensional ITG structures are formed near the boundary of marginal stability too. This assumption is supported by numerical simulations of the similar nonlinear system describing ITG two-dimensional structures formation on nonlinear stage of instability [1]. The nonlinear instability which is associated with an existence of non-positively defined energy invariant developed. The Hamiltonian of the system describing modified explosive instability of zero-energy modes is also not positively defined [18,19]. We believe that our consideration makes clearer the mechanisms leading to drift structures formation near the boundary of marginal stability.

REFERENCES

1. Ottaviani, M., Romanelli, F., and Benzi, R., *Phys. Fluids* **B2**, 67-74 (1990).
2. Horton, W., Su, K.N., and Morrison, P.I., *Sov. J. of Plasma Phys.* **16**, 562-568 (1990).
3. Pavlenko, V., and Weiland, J., *Physica Scripta* **47**, 96-98 (1993).
4. Riccardi, C., Gamterale, L., Salierno, M., and Fontanesi, M., "Drift wave destabilization in toroidal plasma," presented at the Int. Conf. on Plasma Physics, Nagoya, Japan, 1996.
5. Weiland, J., *Current Topics in the Phys. Fluids* **1**, 439-460 (1994).
6. Connor, J.W., *Plasma Phys. Control. Fusion* **37**, A119-A133 (1995).
7. Nordman, H., and Weiland, J., *Nucl. Fusion* **29**, 251-263 (1989).
8. Romanelly F., Tang W.H., and White R.B., *Nucl. Fusion* **26**, 1515-1528 (1986).
9. Kurki-Suonio, T.K., Groebner, R.J., and Burrell, K.H., *Nucl. Fusion* **32**, 133-142 (1992).
10. Rewoldt, G., and Tang, W.M., *Phys. Fluids* **30**, 807-817 (1987).
11. Coppi, B., *Comments Plasma Phys. Control. Fusion* **5**, 261-269 (1980).
12. Terry, P.W., Leboeuf. J.N., Diamond, P.H., Thayer, D.R., Sedlak, J.E., and Lee, G.S., *Phys. Fluids* **31**, 2920-2927 (1988).
13. Kishimoto, Y., Tajima, T., Horton, W., LeBrun, M.J., and Kim, J.-Y., *Phys. Plasmas* **3**, 1289-1307 (1996).
14. Dauchot, O., and Daviaud F., *Phys. Fluids* **A7**, 901-903 (1995).
15. Walts, R.S., *Phys. Rev. Lett.* **55**, 1098-1101 (1985).

16. Nordman, H., Pavlenko, V.P., and Weiland, J., *Phys. Fluids* **B5**, 402-408 (1993).
17. Davydova, T.A., Pavlenko, V.P., Taranov, V.B., and Shamrai, K.P., *Plasma Phys.* **20**, 333-381 (1978).
18. Davydova, T.A., "Explosive instability of "zero-energy waves" and its stabilization in a plasma with magnetized ion beam" in *Proc. Int. Conf. on Plasma Phys.*, Göteborg, 1982, p.192-195.
19. Davydova, T.A., and Zmudskii, A.A., *Plasma Phys. Reports* **20**, 723-733 (1994).
20. Davydova, T.A., and Oraevskii, V.N., *Zhurnal Eksperimental'noy i Teoreticheskoi Fiziki* **66**, 1613-1621 (1974).
21. Mordovskaya, V.G., and Oraevskii, V.N., *Fizika Plazmy* **11**, 1350-1357 (1985).
22. Davydova, T.A., "Nonlinear and Turbulent Processes in Physics", New York: Harwood Academy Press, 1984, vol. 1, pp. 177-181.
23. Biglary, H., Diamond, P.H., and Rosenbluth, M.N., *Phys. Fluids* **B1**, 109-118 (1989).
24. Romanelli, F., *Phys. Fluids* **B1**, 1018-1025 (1989).
25. Romanelli, F., and Zonca, F., *Phys. Fluids* **B5**, 4081-4089 (1993).
26. Taylor, J.B., Wilson, H.R., and Connor, J.W., *Plasma Phys. Control. Fusion* **38**, 243-250 (1996).
27. Horton, W., Choi, D.I., and Tang, W.M., *Phys. Fluids* **24**, 1077-1085 (1981).
28. Hahm, T.S., and Tang, W.M., *Phys. Fluids* **B1**, 1185-1192 (1989).
29. Dum, C.T., and Ott, E., *Plasma Phys.* **13**, 177-187 (1971).
30. Nakamura, S., Yuyma, T., Kubo, H., and Mitani, K., *J. Phys. Soc. Jpn.* **48**, 2112-2118 (1980).
31. Yamagiwa, K., Tokuda, K., and Mieno, T., in *Proc. Int. Conf. on Plasma Phys.*, New Dehli, India, 1989, vol. 3, pp. 1021-1024.
32. Davydova, T.A., and Lashkin, V.M., *Ukrainskii Fizicheskii Zhurnal* **37**, 1833-1839 (1992).
33. Itoh, K., Itoh, S.-I., Fukuyama A., and Yagi M., *Plasma Phys. Rep.* **22**, 721-739 (1996).
34. Yagi, M., Itoh, S.I., Itoh, K., Fukuyama, A., and Azumi, M., *Phys. Plasmas* **2**, 4140-4148 (1995).
35. Oraevskii, V.N., Tasso, H., and Wobig, H., in *Proc. of the Int. Conf. on Plasma Phys. and Control. Nuclear Fusion Research*, IAEA, Vienna, vol. 1, pp. 671-674.
36. Petviashvili, V.I., *Sov. J. Plasma Phys.* **3**, 151-154 (1977).
37. Pokhotelov, O.A., Stenflo, L., and Shukla P.K., *Plasma Phys. Rep.* **22**, 941-953 (1996).

Kinetics of Muon Longitudinal Cooling

Zohreh Parsa[†1] and Pavel Zenkevich[‡]

†Department of Physics, Bldg. 901A
Brookhaven National Laboratory
PO Box 5000 Upton, NY 11973-5000

‡Institute for Theoretical and Experimental Physics
B Cheremushkinskaya Ulitsa 25, RU-117 259, Moscow, Russia

Abstract. Kinetics of longitudinal ionization cooling is analysed by use of method of moments. Special attention is given to "supercooling" situation when longitudinal damping coefficient is much more than synchrotron frequency. It is shown that in this case the emittance damping becomes slower, and it should be taken into account in designing of the ionization cooling system.

INDIVIDUAL PARTICLE OSCILLATIONS

For muon muon collider [1] a longitudinal ionization is created by "wedges," which represent the pieces of matter with a transverse gradient of the "ionization losses"; these pieces should be placed at sections where dispersion exisits. Inside such pieces change of Δp (deviation of a muon momentum p from its equalibrium value p_s) is proportional to Δp, which results in longitudinal damping. Equation of longitudinal motion for individual particle in a presence of wedges may be written in the following form:

$$y'' + 2\alpha y' + \Omega_s^2 y = W_y(z). \qquad (1)$$

Here $y = z - z_s$ (z is a longitudinal coordinate of particle, z_s is a longitudinal coordinate of equilibrium particle), 2α is a longitudinal damping coefficient, Ω_s is a synchrotron frequency, $W_y(z)$ describes the heating effect due to "knock-on" electrons.

A longitudinal damping coefficient is defined by the following formula:

---

[1]) Work performed under the auspices of the U.S. Department of Energy contract no. DE-AC02-76CH00016, and under National Science Foundation grant no. PHY94-07194.

$$2\alpha = \frac{1}{mc^2\beta^2\gamma} \left\langle \left(\frac{dE}{dz}\right)_{\text{ion}} \frac{\Psi}{L_0} \right\rangle + \frac{\frac{d}{dz}(p/\Lambda)}{p/\Lambda} - \frac{d}{dp}\left[\left(\frac{dE}{dz}\right)_{\text{ion}} \frac{1}{v}\right] \quad (2)$$

Here: β, γ are relativistic factors; m is a muon mass; sign $\langle \rangle$ means averaging on z; $\left(\frac{dE}{dz}\right)_{\text{ion}}$ are ionization losses of energy per unit of length, depending on a transverse coordinate x; Ψ is the dispersion function. Parameter L_0 is defined by the following expression:

$$L_0 = \frac{\left(\frac{dE}{dz}\right)_{\text{ion}}}{\frac{d}{dx}\left(\frac{dE}{dz}\right)_{\text{ion}}}; \quad (3)$$

Momentum deviation Δp is connected to our variable y by standard expressions:

$$\Lambda \frac{\Delta p}{p} = \frac{dy}{dz} \quad (4)$$

$$\Lambda = \left\langle \frac{\Psi}{R} \right\rangle + \frac{1}{\gamma^2} \quad (5)$$

Emittance in our variables is defined by

$$\epsilon_y^2 = \langle (y')^2 \rangle \langle y^2 \rangle - \langle yy' \rangle^2, \quad (6)$$

and a normalized longitudinal emittance is given by

$$\epsilon_N^2 = \left[\langle (\Delta p)^2 \rangle \langle y^2 \rangle - \langle \Delta p \cdot y \rangle^2\right] \frac{1}{(mc)^2} \quad (7)$$

Using Eq. (4), we find:

$$\epsilon_N = \frac{\beta\gamma}{\Lambda} \epsilon_y \quad (8)$$

The longitudinal (synchrotron) frequency Ω is defined by:

$$\Omega^2 = \frac{2\pi e E_{ac} \cdot \cos\left(\frac{2\pi}{\lambda} z_s\right) \Lambda}{mc^2 \lambda \beta^2 \gamma} \quad (9)$$

Here E_{ac} is an amplitude of the r.f. field, and λ is its wavelength.

Let us consider a case when α and Ω are constant. Eigenvalues of Eq. (1) are defined by:

$$\lambda_{1,2} = -\alpha \pm \sqrt{\alpha^2 - \Omega^2} \quad (10)$$

We have two different regimes of operation:

a) $\alpha > \Omega$, non-periodic motion;
b) $\alpha < \Omega$, periodic oscillations.

At the first regime both modes are damped with different damping rates. For $\alpha \gg \Omega$ a damping rate of "slow" mode is defined by:

$$\lambda_1 = (\alpha - \sqrt{\alpha^2 - \Omega^2}) = -\Omega^2/2\alpha \qquad (11)$$

For optimal cooling it is necessary that $\alpha = \alpha_\perp$ where $\alpha_\perp$ is a transverse damping coefficient.

Numerical estimations show that in this case for reasonable amplitudes of the accelerating field $\alpha \gg \Omega$, and a damping rate of "slow" mode is defined by Eq. (11).

For periodic oscillations

$$\lambda_{1,2} = -\alpha \pm i\sqrt{\Omega^2 - \alpha^2} = -\alpha \pm iw_1 \qquad (12)$$

We see that both waves are damped with the same damping rate.

Let $y(z=0) = y_0, y'(z=0) = y'_0$; then we can write:

$$y = \frac{\lambda_2 \varphi_1 - \lambda_1 \varphi_2}{\lambda_2 - \lambda_1} y_0 + \frac{\varphi_2 - \varphi_1}{\lambda_2 - \lambda_1} y'_0$$
$$y' = \Omega^2 \frac{\varphi_2 - \varphi_1}{\lambda_2 - \lambda_1} y_0 + \frac{\lambda_2 \varphi_2 - \lambda_1 \varphi_1}{\lambda_2 - \lambda_1} y'_0 \qquad (13)$$

Here $\varphi_{1,2} = \exp(\lambda_{1,2} z)$. For aperiodic case:

$$\frac{\lambda_2 \varphi_1 - \lambda_1 \varphi_2}{\lambda_2 - \lambda_1} = \exp(-\alpha z)\left[ch\alpha_0 z + \frac{\alpha}{\alpha_0} sh\alpha_0 z\right] \qquad (14)$$

$$\frac{\varphi_2 - \varphi_1}{\lambda_2 - \lambda_1} = \frac{\alpha \exp(-\alpha z)}{\alpha_0} sh\alpha_0 z \qquad (15)$$

$$\frac{\lambda_2 \varphi_2 - \lambda_1 \varphi_1}{\lambda_2 - \lambda_1} = \exp(-\alpha z)\left[ch\alpha_0 z - \frac{\alpha}{\alpha_0} sh\alpha_0 z\right] \qquad (16)$$

We can obtain similar expressions for periodic case from Eqs. (14)–(16) using change:

$$\left.\begin{array}{c} ch\alpha_0 z \to \cos w_1 z \\ \frac{sh\alpha_0 z}{\alpha_0} = \frac{\sin w_1 z}{w_1} \end{array}\right\} \quad w_1 = \sqrt{\Omega^2 - \alpha^2} \qquad (17)$$

Using Eqs. (14)–(16) it is easy to calculate an evolution of moments and the emittance by direct calculations. However, in order to take into account the heating it is simpler to use method of moments.

EQUATIONS FOR MOMENTS AND ITS SOLUTION

Using Eq. (1), we derive a system of the first order equations for moments $\langle y^2 \rangle$; $\langle y \frac{dy}{dz} \rangle$ and $\langle \left(\frac{dy}{dz}\right)^2 \rangle$. Let us define: $\langle y^2 \rangle = u$; $\langle y \frac{dy}{dz} \rangle = v$, $\langle \left(\frac{dy}{dz}\right)^2 \rangle = t$. Then we can write

$$\frac{du}{dz} = 2v \tag{18}$$

$$\frac{dv}{dz} = t - 2\alpha v - \Omega^2 u \tag{19}$$

$$\frac{dt}{dz} = -2\Omega^2 v - 4\alpha t + W_y \tag{20}$$

$$\frac{d\epsilon_y^2}{dz} = -4\alpha \epsilon_y^2 + tW_y \tag{21}$$

It is interesting to compare Eq. (20) with "standard equation" for $\langle (\Delta p)^2 \rangle$ which is usually used in literature [1,2]. Using Eq. (4), we obtain:

$$\frac{d}{dz} \langle (\Delta p)^2 \rangle = -4\alpha \langle (\Delta p)^2 \rangle - 2\Omega \frac{2\Lambda}{p} \langle y \cdot \Delta p \rangle + W_p \tag{22}$$

We see that this equation differs from the standard one by a presence of the additional second term in RHS. Really, this term does not have affect on equilibrium parameters; however, this term has to be taken into account for calculation of evolution of the moments and the emittance in time.

Let us consider a case when α, Ω are constant. Then our system of four linear equations with constant coefficients can be solved analytically.

Equilibrium solution of the system can be found, if we assume that all derivatives are equal to zero. Then (for α, Ω, W_y are constant) we obtain

$$\langle y^2 \rangle_{eq} = W_y / 4\alpha \Omega^2 \tag{23}$$

$$\left\langle \left(\frac{dy}{dz}\right)^2 \right\rangle_{eq} = W_y / 4\alpha \tag{24}$$

$$\epsilon_y^{eq} = W_y / 4\alpha \Omega \tag{25}$$

We see that for $\Omega = 0$ an equilibrium solution is absent.

If at an initial point the beam has a longitudinal crossover, initial conditions may be written as follows:

$$\left. \begin{array}{ll} u(z=0) = u_0; & v(z=0) = 0; \\ t(z=0) = t_0; & \epsilon(z=0) = \epsilon_0 \end{array} \right\} \tag{26}$$

We can solve our system with these initial conditions by use of Laplace transform (see Appendix).

In a case of "superdamping" ($\alpha \gg \Omega$) it is possible to obtain comparatively simple asymptotic expressions for all interesting parameters. Let us assume that $x = \alpha/\Omega$, $\tau = \Omega z$, and that the initial beam is "matched," i.e., $t_0 = \Omega^2 u_0$. Then, using Eqs. (45)–(47) we obtain (for $exp(-2x\tau) \ll 1$):

$$\frac{u_{fr}}{u_0} = \frac{1}{2} exp(-\tau/x) \frac{x}{\sqrt{x^2-1}} \left[\frac{x}{\sqrt{x^2-1}} - 1\right] \tag{27}$$

$$\frac{v_{fr}}{u_0} = -\frac{1}{2} \Omega exp(-\tau/x) \frac{x}{x^2-1} \tag{28}$$

$$\frac{t_{fr}}{u_0} = \frac{1}{2} \Omega^2 exp(-\tau/x) \frac{x}{\sqrt{x^2-1}} \left[\frac{x}{\sqrt{x^2-1}} - 1\right] \tag{29}$$

where u_{fr}, v_{fr} and t_{fr} correspond to free oscillations (for $w_y = 0$). It is easy to see, that

$$u_{fr} \cdot t_{fr} - v_{fr}^2 = 0 \tag{30}$$

Thus, we observe the following mechanism of the longitudinal free emittance damping in "superdamping" regime: a) the beam size is damped very slowly and practically doesn't change (see Eq. (27)); b) the beam spread on momenta is fast decreased to asymptotic value:

$$\frac{t}{t_0} = \left[\frac{1}{2} \frac{x}{\sqrt{x^2-1}} \left(\frac{x}{\sqrt{x^2-1}} - 1\right)\right]^{1/2} \tag{31}$$

and then it is diminished very slowly.

Attenuation of the emittance is due to the "crossing" moment term $\left(\langle y\frac{dy}{dz}\rangle\right)$.

In the region of slowly changed t, value of ϵ_y^2 is determined to high accuracy by the following formula:

$$\epsilon_y^2 \simeq \frac{u(z) \cdot W_y}{4\alpha} \tag{32}$$

Thus, in order to find $u(z)$ we must add to u_{fr} an expression for u_{heat}. Using Eq. (48), we find:

$$\frac{u_{heat}}{u_{eq}} \simeq 1 - \frac{x \cdot (x + \sqrt{x^2-1})}{(x^2-1)2} exp(-\tau/x) \tag{33}$$

Using Eq. (27) and (32)–(33) we obtain the final result:

$$\epsilon_y \simeq \epsilon_{eq} \left[1 + \left(\frac{u_0}{u_{eq}} - 1\right) exp\left(-\frac{\tau}{x}\right)\right]^{1/2} \tag{34}$$

Usually $\frac{u_0}{u_{eq}} = \frac{\epsilon_0}{\epsilon_{eq}} \gg 1$, and with good accuracy we can write:

$$\epsilon_y \simeq (\epsilon_{eq} \cdot \epsilon_0)^{1/2} exp(-\tau/2x) \qquad (35)$$

It is easy to see that for normalized emittances Eq. (35) is also valid. For normalized equilibrium emittance we have the following expression:

$$\epsilon_{eq} = \frac{1}{4\alpha \cdot \Omega} \cdot \frac{d}{dz} \frac{\langle(\Delta p)^2\rangle}{(mc)} = \frac{1}{4\alpha \cdot \Omega \beta^2 (mc^2)} \frac{d}{dz} \langle(\Delta E)^2\rangle \qquad (36)$$

Here

$$\frac{d}{dz}\langle(\Delta E)^2\rangle = K_s \gamma^2 \left(1 - \frac{\beta^2}{2}\right) \qquad (37)$$

$$K_s = 4\pi (r_e m_e c^2)^2 \frac{\rho Z}{A} \qquad (38)$$

Here r_e, m_e - classical radius and mass of electron; m - muon mass ; ρ, Z, A are density, charge and atomic number of the wedge material.

DISCUSSION

We see that kinetics of the longitudinal ionization cooling for strong damping coefficient have a lot of special features:
- Longitudinal oscillations of individual particles have nonperiodic character with different damping rate of two modes of oscillations.
- During the cooling longitudinal beam becomes "unmatched."
- Longitudinal beam size is changing very slowly, and therefore time necessary to reach the equilibrium emittance is increased with an enhancement of the longitudinal damping rate.

It is clear that these special features of longitudinal cooling should be taken into account when we look for the optimal characteristics of the cooling system.

APPENDIX

Let us apply Laplace transform to Eqs. (19)–(21) with inital conditions (26). Thus we obtain

$$\left. \begin{array}{r} pU - 2V = u_0 \\ \Omega^2 U + (p + 2\alpha)V - T = 0 \\ 2\Omega^2 V + (p + 4\alpha)T = \frac{w}{p} + t_0 \end{array} \right\} \qquad (39)$$

For emittance we have the following equation

$$(p+4\alpha)\Phi = \Phi_0 + wU(p) \qquad (40)$$

Solving this linear system, we have:

$$U(p) = \frac{2(\Omega^2 u_0 + t_0) + u_0(p+2\alpha)(p+4\alpha)}{\Delta_0(p)} + \frac{2w}{p\Delta_0(p)} \qquad (41)$$

$$V(p) = \frac{p(t_0 - \Omega^2 u_0) - 4\alpha u_0 \Omega^2}{\Delta_0(p)} + \frac{w}{\Delta_0(p)} \qquad (42)$$

$$T(p) = \frac{2\Omega^4 u_0 + t_0[p(p+2\alpha) + 2\Omega^2]}{\Delta_0(p)} + \frac{w[p(p+2\alpha) + 2\Omega^2]}{p\Delta_0(p)} \qquad (43)$$

Denominator $\Delta_0(p)$ is given by $\Delta_0(p) = (p+2\alpha)(p^2 + 4\alpha p + 4\Omega^2)$.

The first terms in Eqs. (41)–(43) describe evolution of free oscillations, the second ones - evolution due to heating. Substituting Eq. (42) in Eq. (41), we obtain:

$$\Phi = \frac{\Phi_0}{p+4\alpha} + \frac{w[2(\Omega^2 u_0 + t_0) + u_0(p+2\alpha)(p+4\alpha)]}{\Delta_0(p)(p+4\alpha)} + \frac{2w^2}{p(p+4\alpha)\Delta_0(p)} \qquad (44)$$

Here U, V, T and Φ are, correspondingly, Laplace transforms for u, v, t and ϵ^2. The first term describes a damping of the initial emittance, the second term describes the emittance growth due to synergism between free oscillations and the heating, the last term describes the emittance growth due to heating in an absence of an initial phase volume.

Using the backward Laplace transform, we obtain for terms, describing a damping of free oscillations:

$$u_{fr}(z) = exp(-2\alpha z)\left\{u_0\left(ch2\alpha_0 z + \frac{\alpha}{\alpha_0}sh2\alpha_0 z\right) + \frac{\Omega^2 u_0 + t_0}{2\alpha_0^2}(ch2\alpha_0 z - 1)\right\} \qquad (45)$$

$$v_{fr}(z) = exp(-2\alpha z)\left\{\frac{\alpha(t_0 + \Omega^2 u_0)}{2\alpha_0^2}(1 - ch2\alpha_0 z) + \frac{t_0 - \Omega^2 u_0}{2\alpha_0}sh2\alpha_0 z\right\} \qquad (46)$$

$$t_{fr}(z) = exp(-2\alpha z)\left\{\frac{\Omega^2(t_0 + \Omega^2 u_0)}{2\alpha_0^2}(ch2\alpha_0 z - 1) + t_0\left(ch2\alpha_0 z - \frac{\alpha}{\alpha_0}sh2\alpha_0 z\right)\right\} \qquad (47)$$

For the second terms (due to heating) we obtain:

$$\frac{u_{\text{heat}}(z)}{u_{eq}} = 1 + \frac{\Omega^2}{\alpha_0^2}exp(-2\alpha z) - \frac{\alpha exp(-2\alpha z)}{\alpha_0^2} \cdot (\alpha_0 sh2\alpha_0 z + \alpha ch2\alpha_0 z) \quad (48)$$

$$v_{\text{heat}}(z) = \frac{w}{4\alpha_0^2}exp(-2\alpha z)(ch2\alpha_0 z - 1) \quad (49)$$

$$\frac{t_{\text{heat}}(z)}{t_{eq}} = 1 + exp(-2\alpha z)sh2\alpha_0 z\left(1 - \frac{\alpha}{\alpha_0}\right) - exp(-2\alpha z)\frac{\alpha^2 ch2\alpha_0 z - \Omega^2}{\alpha_0^2} \quad (50)$$

It is easy to see, that $\frac{u_{\text{heat}}}{u_{eq}}$ and $\frac{t_{\text{heat}}}{t_{eq}} \to 1$ for $z \to \infty$; $v_{\text{heat}} \to 0$ for $z \to \infty$. Expresion for $\epsilon^2(z)$ have the more complicated form:

$$\epsilon^2_{(z)} = \epsilon_0^2 exp(-4\alpha z) + \frac{w(\Omega^2 u_0 + t_0)}{4}$$
$$\left\{-\frac{1}{4\alpha}\left[\frac{exp(-4\alpha z)}{\Omega^2} + \frac{exp(-2\alpha z)}{\alpha_0^2}\right] + \frac{exp(-2\alpha z)}{2\alpha_0^2 \Omega^2}[\alpha ch2\alpha_0 z - \alpha_0 sh2\alpha_0 z] + \right.$$
$$\frac{wu_0}{\alpha_0}exp(-2\alpha z)sh2\alpha_0 z +$$
$$\left. \epsilon_{eq}^2\left[1 + exp(-4\alpha z) - 2exp(-2\alpha z)\frac{\alpha^2 ch2\alpha_0 z - \Omega^2}{\alpha_0^2}\right]\right\} \quad (51)$$

Here $\alpha_0 = \sqrt{\alpha^2 - \Omega^2}$. Expressions for oscillation regime can be found by substitutions using:

$$\left.\begin{array}{rcl}ch2\alpha_0 z & \to & \cos 2w_1 z \\ \frac{sh2\alpha_0 z}{\alpha_0} & \to & \frac{\sin 2w_1 z}{w_1}\end{array}\right\} w_1 = \sqrt{\Omega^2 - \alpha^2} \quad (52)$$

ACKNOWLEDGEMENTS

This work was partially carried out at the Institute for Theoretical Physics, University of California, Santa Barbara, as part of the New Ideas for Particle Accelerators Program with partial support by the the National Science Foundation Grant No. NSF-PHY-94-07194.

REFERENCES

1. "$\mu^+\mu^-$ Collider, A Feasibility Study," BNL-52503, Fermilab-Conf-961092; LBLNL-38946, Snowmass, July 1996.
2. T.A. Vsevolozhskaya, "Kinetics of Ionization Cooling of Muons," Beam Dynamics and Technology Issues for $\mu^+\mu^-$ Collider, (1996) p. 159.

Decoherence and Wakefield Effects in One-Turn Map Measurement

Chun-xi Wang and John Irwin

*Stanford Linear Accelerator Center,
Stanford University, Stanford, CA 94309 USA*

Abstract. To measure the one-turn map of a storage ring, a particle beam is used to explore the single particle dynamics. Since the behavior of a beam is quite different from that of a single particle, it is important to evaluate the differences and their effects on measurements. In this paper, we discuss two major physical effects: decoherence and wakefield effects. Our purpose is to explore the criteria, under which these effects are negligible in map measurements.

INTRODUCTION

Due to the importance of the one-turn map in studies of nonlinear beam dynamics of a storage ring and the possibility to measure such a map with present technology [1], serious efforts are being made to carry out such measurement at SLAC. The measurement procedure is quite straightforward in principle. We kick the beam to various phase space locations and collect the turn-by-turn beam positions in phase space. Two successive positions form a pair of data which are connected via the one-turn map. A large number of data pairs with sufficient sampling of the phase space volume are collected and fit to a Taylor map.

Obviously, the accuracy requirement of the data is very high due to the weak nonlinearities we are measuring. Many factors limit the measurement accuracy. In this paper, we deal with two major physical factors originating from the difference between beam and single particle. The first one is beam decoherence. It is well-known that the coherent motion of a kicked beam will be significantly changed because of the tiny tune spread among particles in a bunch. The other one is collective effects due to wakefields. We will only discuss the short-range wakefield effect in this paper. Since only a single bunch, low current and a relatively small number of turns are used in map measurement, the long-range wakefields due to previous turns are expected to be much weaker.

Our goal is to find the conditions under which these effects are negligible for a given accuracy requirement. Using analytical results of a simplified decoherence model, we examine the deviation of the beam centroid from single particle motion and compare it with BPM resolution. We then address the concept of the beam-centroid map and discuss decoherence effects via the centroid map. For the wakefield effects, we compute the wake force on the beam centroid and compare it with nonlinear lattice forces.

DECOHERENCE BASED ON A SIMPLE MODEL

For a normal operation ring, the one-turn map generally has very weak nonlinearities and the tune is placed away from significant resonances. Therefore, it is safe to assume (in a good approximation) that the map has a non-resonant normal from. In another words, particle motion can be canonically transformed into a circular phase-space motion with a tune depending on its energy and action as determined by chromaticity and tune-shift-with-amplitude respectively:

$$\nu_x = \nu_x^0 + \xi_x \delta + a_x J_x + a_{xy} J_y + \cdots \tag{1}$$

Due to such dependencies and the natural energy spread as well as emittance of a beam, the motions of particles in a kicked beam will decohere so that the centroid motion is very different from that of a single particle.

The centroid motion of a kicked beam has been calculated in many works [2–6]. Here we give a generalized result whose computation is similar to the references.

$$\langle x \rangle \simeq x_k \sqrt{1 + \alpha_x^2}\, F\, \bar{A} \cos(2\pi \nu_x^0 n + \Delta\bar{\phi} - \arctan \alpha_x) \tag{2}$$

where

$$F \equiv \exp\left[-2(\frac{\xi_x \sigma_\delta}{\nu_s})^2 \sin^2(\pi \nu_s n)\right]$$

$$\bar{A} \equiv \frac{1}{1+\theta_x^2} \exp\left[-\frac{k_x^2}{2}\frac{\theta_x^2}{1+\theta_x^2}\right] \cdot \frac{1}{\sqrt{1+\theta_{xy}^2}} \exp\left[-\frac{k_y^2}{2}\frac{\theta_{xy}^2}{1+\theta_{xy}^2}\right]$$

$$\Delta\bar{\phi} \equiv \frac{k_x^2}{2}\frac{\theta_x}{1+\theta_x^2} + 2\arctan\theta_x + \frac{k_y^2}{2}\frac{\theta_{xy}}{1+\theta_{xy}^2} + \arctan\theta_{xy}$$

and $\theta_x \equiv 2\pi a_x \epsilon_x n$, $\theta_{xy} \equiv 2\pi a_{xy} \epsilon_y n$. $\epsilon_{x,y}$ are beam emittance and n is the number of turns after the kick. $k \equiv \sqrt{1+\alpha^2}\frac{x_k}{\sigma} = \sqrt{\gamma x_k^2/\epsilon}$ is a dimensionless quantity that measures the effective kick strength in terms of the initial beam size σ. α_x and γ_x are Twiss parameters at the measurement location. We assumed no linear coupling among horizontal, vertical, and longitudinal phase spaces. In the following we first discuss decoherence effects due to nonlinearities, then examine the effect due to chromaticity decoherence.

The $\epsilon \to 0$ limit of Eq.(2) is the corresponding single particle motion, where $F = \bar{A} = 1$ and the phase shift $\Delta\bar{\phi}_0 = (k_x^2 \theta_x + k_y^2 \theta_y)/2$. The deviation of beam centroid from the single particle motion is simply:

$$\left|\frac{\Delta x}{x_{max}}\right|(n) = \left|2\sin\frac{\Delta\bar{\phi} - \Delta\bar{\phi}_0}{2}\sin(\phi + \frac{\Delta\bar{\phi} + \Delta\bar{\phi}_0}{2}) + (1-\bar{A})\cos(\phi+\Delta\bar{\phi})\right|$$

$$\simeq \left|(2\hat{\theta}_x + \hat{\theta}_{xy})\sin\phi\right| n + O(\theta^2) \qquad (3)$$

where the unperturbed phase $\phi = 2\pi\nu_x^0 n - \arctan\alpha_x$ and constant $\hat{\theta} \equiv \theta/n$. Note that, in the leading order, the deviation increases linearly with the number of turns and is independent of the kick strength. Compared with simulations done with all nonlinearities in a ring, the simple result in Eq.(3) gives a reasonablely good expression of the deviation.

The simplest way to eliminate decoherence effects on map measurements is to limit the number of turns used for data collection after each kick according to:

$$n < \left|2\hat{\theta}_x + \hat{\theta}_{xy}\right|^{-1} \frac{\text{BPM resolution}}{x_{max}} \qquad (4)$$

A similar condition can be set for vertical plane. To get a feeling of how strong this requirement is, we consider the PEP-II high energy ring [7] as an example. For the design lattice, $\epsilon_x = 5 \times 10^{-8}$, $\epsilon_y = 2 \times 10^{-9}$, $a_x = 110$, and $a_{xy} = 1940$. Thus $2\hat{\theta}_x + \hat{\theta}_{xy} \simeq 10^{-4}$. If we require a signal-to-noise ratio of 10^3 (e.g. a $10\mu m$ BPM resolution and a 10 mm kick), only 10 turns can be used. From this point of view, decoherence effect is quite significant.

However, one may argue that, Eq.(3) gives the accumulated deviation between particle and beam motion. In map measurement, beam position will be measured turn-by-turn. Therefore, the error at n-th turn due to decoherence is not given by Eq.(3). Instead, we should start with the $(n-1)$-th turn centroid and track it as a single particle, then compare the result with the n-th turn centroid. The difference represents the measurement error we see at n-th turn. To estimate this, we can simply calculate the phase difference as:

$$\frac{\Delta x}{x} \simeq \Delta\bar{\phi}(n) - \left[\Delta\bar{\phi}(n-1) + \frac{k_x^2}{2}\hat{\theta}_x \bar{A}_x^2(n-1) + \frac{k_y^2}{2}\hat{\theta}_{xy} \bar{A}_y^2(n-1)\right]$$

$$\simeq \frac{k_x^2}{2}\hat{\theta}_x\left[1-\bar{A}_x^2(n)\right] + \frac{k_y^2}{2}\hat{\theta}_{xy}\left[1-\bar{A}_y^2(n)\right]$$

$$\simeq \frac{1}{2}\left[k_x^4\hat{\theta}_x^3 + k_y^4\hat{\theta}_{xy}^3 + k_x^2 k_y^2(\hat{\theta}_x\hat{\theta}_{xy}^2 + \hat{\theta}_y\hat{\theta}_{yx}^2)\right] n^2 \qquad (\text{if } k\hat{\theta}n < 1) \qquad (5)$$

where $\bar{A}$ is assumed to take into account the single particle tune-shift at the damped amplitude. Take the above example again and consider a 8σ horizontal kick, one gets $n \simeq 3700$. According to this argument, decoherence should not be a concern at all in this case. Although this exercise reveals that the criterion in Eq.(4) may significantly overestimate the decoherence effects, we should not use Eq.(5) as a criterion because of the approximate nature of Eq.(2). In fact, numerical simulation indicates that the correct answer for our example is a couple of hundred turns instead of thousands of turns.

Now we take a look at the chromaticity decoherence effect. For simplicity, we consider just the change of factor F in a turn.

$$\frac{\Delta x}{x} \simeq 2(\frac{\xi_x \sigma_\delta}{\nu_s})^2 \left[\sin^2(\pi\nu_s n) - \sin^2 \pi\nu_s(n-1)\right]$$

$$\simeq 2\pi \frac{(\xi_x \sigma_\delta)^2}{\nu_s} \sin(2\pi\nu_s n) \qquad (6)$$

For our high energy ring example, $\sigma_\delta = 6 \times 10^{-4}$ and $\nu_s = 0.052$. Take $\xi_x \simeq 1$, the maximum deviation is 4×10^{-5}, which is much less than the BPM sensitivity 10^{-3}. However, if the chromaticity is much larger, the maximum deviation could become large due to the square dependency. In such a case, one may have to rely on the factor $\sin(2\pi\nu_s n)$ to reduce the error, i.e. only use a few turns of data after a kick.

The problem with the analytical approach is that the simple decoherence model used in Eq.(2) may not be adequate. The actual behavior of the beam centroid is much more complex. Numerical simulation is necessary to obtain a more reliable prediction for a given lattice. Such simulation is often very time consuming because a large number of particles have to be tracked to obtain a sufficiently accurate beam centroid. In the next section, we examine the decoherence effects from another perspective.

ONE-TURN BEAM-CENTROID MAP

In map measurement, we are interested in the single-particle map

$$X_k^f = X_k^0 + R_{kl}X_l^i + T_{klm}X_l^i X_m^i + U_{klmp}X_l^i X_m^i X_p^i + \cdots \qquad (7)$$

where the summation convention on the repeated indices is assumed. X_k is the k-th component of a phase space vector. Every particle in a beam follows the same single-particle map. Therefore, the beam centroid $\bar{X}$ follows

$$\bar{X}_k^f = [X_k^0 + \hat{T}_{klm}\langle\delta X_l^i \delta X_m^i\rangle + \hat{U}_{klmp}\langle\delta X_l^i \delta X_m^i \delta X_p^i\rangle + \cdots]$$
$$+ \bar{X}_l^i[R_{kl} + \hat{U}_{klmp}\langle\delta X_m^i \delta X_p^i\rangle + \hat{V}_{klmpq}\langle\delta X_m^i \delta X_p^i \delta X_q^i\rangle + \cdots]$$
$$+ \bar{X}_l^i \bar{X}_m^i[T_{klm} + \hat{V}_{klmpq}\langle\delta X_p^i \delta X_q^i\rangle + \cdots]$$
$$+ \cdots$$
$$\equiv \bar{X}_k^0 + \bar{R}_{kl}\bar{X}_l^i + \bar{T}_{klm}\bar{X}_l^i \bar{X}_m^i + \bar{U}_{klmp}\bar{X}_l^i \bar{X}_m^i \bar{X}_p^i + \cdots \qquad (8)$$

where $\langle \cdots \rangle$ means an average over beam phase space distribution of X^i (assumed symmetric) and $\delta X = X - \bar{X}$ is a particle's deviation from the centroid.[1] The hatted coefficients are related to the unhatted ones by constant factors $\prod_i \binom{n(i)}{m(i)}$, where $\binom{n}{m}$ is the binomial coefficient and, $n(i)$ and $m(i)$ are the number of occurrences of certain index i among the running indices in the coefficients and in the $\langle \cdots \rangle$ terms respectively.[2] Note that permutations of the running indices is assumed not to contribute in Eqs. (7) and (8). We see that the beam centroid does not follow the single-particle map. Extra terms appear and depend on the beam phase space distribution. The difference between the centroid map and single-particle map decreases with the beam emittance. The two maps become the same if the nonlinearity is negligible.

It is important to notice that, as long as the beam maintains the same distribution, the coefficients $\bar{X}^0$, $\bar{R}$, $\bar{T}$, $\bar{U}$, etc. are constant, i.e. the beam centroid follows a well-defined Taylor map also, although the centroid map is different from the single-particle map. This observation allows one to overcome the single particle vs. beam problem by requiring that the beam phase space distribution should not change so much that the centroid map is changed significantly. According to this criteria, one should examine the change of centroid map with emittance growth due to decoherence. Usually, we compute the beam-centroid map for a damped beam. The difference between the centroid map and the single-particle map is small but measurable, and changes roughly linearly with beam emittance and energy spread. Therefore, from the map differences and the tolerable error in map measurement, one can estimate how much emittance growth is tolerable and therefore how many turns can be used for a map measurement.

WAKEFIELD EFFECTS

To estimate the wakefield effects, we calculate the wake-force kick on the beam centroid when a Gaussian beam passing some axial-symmetric wake structure characterized by a wake function $W_m(z)$. We assume the longitudinal and transverse beam distributions are uncorrelated, but correlation is allowed between the two transverse degree of freedom.

In the ultra-relativistic approximation, the integrated wake-force impulse between two electrons can be represented by

---

[1] For Gaussian distribution with $\langle \delta X \rangle = 0$, it is well known that $\langle \delta X_p \delta X_q \rangle = \sigma_{pq}^2 = \sigma_p \sigma_q \rho_{pq}$, where ρ_{pq} is the covariance matrix. And $\langle \delta X_p^n \rangle = \begin{cases} 0 & n \text{ odd} \\ (n-1)!! \sigma_p^n & n \text{ even} \end{cases}$

[2] It is not important to understand these factors. Nonetheless, here is an example to help. In 5D, the index i runs from 1 to 5. The term $\hat{U}_{1224} \langle \delta X_2^i \delta X_4^i \rangle$ in $\bar{X}_l^i \hat{U}_{1lmp} \langle \delta X_m^i \delta X_p^i \rangle$ of Eq.(8) should read:

$$\hat{U}_{1224} \langle \delta X_2^i \delta X_4^i \rangle = \binom{0}{0}\binom{2}{1}\binom{0}{0}\binom{1}{1}\binom{0}{0} U_{1224} \langle \delta X_2^i \delta X_4^i \rangle = 2 U_{1224} \langle \delta X_2^i \delta X_4^i \rangle.$$

$$\vec{\bar{F}}_{ij} \equiv \int ds \vec{F}_{ij} = -\nabla V_{ij} \qquad (9)$$

via a potential V_{ij} which reads:

$$V_{ij} = e^2 \sum_{m=0}^{\infty} r_i^m r_j^m \cos[m(\theta_i - \theta_j)] W_m(z_i - z_j) \qquad (10)$$

The impact on the i-th particle due to a N-particle bunch reads:

$$\vec{\bar{F}}_i = -\nabla_{\vec{x}_i} V_i \qquad (11)$$

and

$$V_i(\vec{x}_i) = \sum_{j=1}^{N} V_{ij} = N \int d^3\vec{x}_j \rho(\vec{x}_j) V_{ij}$$

$$= Ne^2 \sum_{m=0}^{\infty} \int d^2\vec{x}_j \rho_\perp(\vec{x}_j) r_i^m r_j^m \Re[e^{im(\theta_i-\theta_j)}] \int_{z_i}^{\infty} dz_j \rho_\parallel(z_j) W_m(z_i - z_j)$$

$$= e \sum_{m=0}^{\infty} \Re[r_i^m e^{im\theta_i} I_m^*] \int_0^{\infty} \frac{d\omega}{\pi} \Im[e^{i\omega z_i/c} \tilde{\rho}_\parallel(\omega) Z_m^\perp(\omega)] \qquad (12)$$

where $\tilde{\rho}(\omega) = \int_{-\infty}^{\infty} dz e^{-i\omega z/c} \rho(z)$ and $Z_m^\perp$ is the transverse impedance. All distributions are normalized to 1, and $\rho = \rho_\perp \rho_\parallel$. The complex m-th moment of the beam is defined as

$$I_m \equiv Ne \int d^2\vec{x} \rho_\perp(\vec{x}) r^m e^{im\theta} \qquad (13)$$

The average longitudinal impact on the bunch centroid is:

$$\langle \vec{\bar{F}}_\parallel \rangle = \frac{1}{N} \sum_{i=1}^{N} -\frac{\partial}{\partial z_i} V_i$$

$$= -e \sum_{m=0}^{\infty} \int d^2\vec{x}_i \rho_\perp(\vec{x}_i) \Re[r_i^m e^{im\theta_i} I_m^*] \int_{-\infty}^{\infty} dz_i \rho_\parallel(z_i) \int_{z_i}^{\infty} dz_j \rho_\parallel(z_j) W_m'(z_i - z_j)$$

$$= -\frac{1}{N} \sum_{m=0}^{\infty} |I_m|^2 \int_0^{\infty} \frac{d\omega}{\pi} |\tilde{\rho}_\parallel(\omega)|^2 \Re[Z_m^\parallel(\omega)]$$

$$= -\frac{1}{N} \sum_{m=0}^{\infty} |I_m|^2 k_l^{(m)} \qquad (14)$$

where $Z_m^\parallel$ is the longitudinal impedance. $k_l^{(m)}$ is often referred as the energy loss factor for the multipole mode m. $N\langle \vec{\bar{F}}_\parallel \rangle$ is known as beam's parasitic energy loss to the impedance.

The average transverse impact on the centroid is:

$$\langle \vec{F}_\perp \rangle = \frac{1}{N} \sum_{i=1}^{N} -(\hat{x}\frac{\partial}{\partial x_i} + \hat{y}\frac{\partial}{\partial y_i})V_i$$

$$= -e \sum_{m=0}^{\infty} \int d^2\vec{x}_i \rho_\perp(\vec{x}_i) \Re[m(r_i e^{i\theta_i})^{m-1}(\hat{x}+i\hat{y})I_m^*] \times$$

$$\int_{-\infty}^{\infty} dz_i \rho_\parallel(z_i) \int_{z_i}^{\infty} dz_j \rho_\parallel(z_j) W_m(z_i - z_j)$$

$$= \frac{1}{N} \sum_{m=0}^{\infty} m(\Re[I_{m-1}I_m^*]\hat{x} - \Im[I_{m-1}I_m^*]\hat{y}) \left[-\int_0^\infty \frac{d\omega}{\pi} |\tilde{\rho}_\parallel(\omega)|^2 \Im[Z_m^\perp(\omega)] \right]$$

$$= \frac{1}{N} \sum_{m=0}^{\infty} m(\Re[I_{m-1}I_m^*]\hat{x} - \Im[I_{m-1}I_m^*]\hat{y}) \, k_\perp^{(m)} \qquad (15)$$

where $k_\perp^{(m)}$ is called transverse kick factor for mode m. Loss factors and kick factors are usually available for the longitudinal monopole mode and transverse dipole mode, which are the dominant modes for a near-axis beam. The following scaling yields a useful rough estimates for higher modes [8].

$$Z_m^\parallel \sim \frac{2}{b^{2m}} Z_0^\parallel \quad \text{and} \quad Z_m^\perp \sim \frac{1}{b^{2m-2}} Z_1^\perp \sim \frac{2\sigma_z}{b^{2m}} Z_0^\parallel \qquad (16)$$

where b is the pipe radius and σ_z is the bunch length.

In the normally used phase space variables, the kicks reads:

$$\Delta\delta = \frac{\langle \vec{F}_\parallel \rangle}{E} \quad \text{and} \quad \Delta\vec{p}_\perp = \frac{\langle \vec{F}_\perp \rangle}{E} \qquad (17)$$

where E is the beam energy.

Up to now, we only assumed there is no correlation between transverse and longitudinal beam distributions. For Gaussian distribution, the calculation can be carried out in detail. The general 2D zero-mean Gaussian distribution in Cartesian and polar coordinate systems are

$$\rho_0(x,y) = \frac{1}{2\pi\sigma_x\sigma_y\sqrt{1-\rho^2}} e^{-\frac{1}{2(1-\rho^2)}(\frac{x^2}{\sigma_x^2} - 2\rho\frac{xy}{\sigma_x\sigma_y} + \frac{y^2}{\sigma_y^2})}$$

$$= \frac{1}{\pi\xi(\sigma_x^2+\sigma_y^2)} e^{-\frac{r^2}{\xi^2(\sigma_x^2+\sigma_y^2)}[1-\sqrt{1-\xi^2}\sin(2\theta+\arctan\frac{\sigma_x^2-\sigma_y^2}{2\rho\sigma_x\sigma_y})]} \qquad (18)$$

where ρ is the correlation coefficient of x,y distribution, $\xi \equiv \frac{2\sigma_x\sigma_y}{\sigma_x^2+\sigma_y^2}\sqrt{1-\rho^2}$.

Using this bunch distribution relative to the beam centroid $\vec{x}_c$, the m-th moment is

$$I_m = Ne \int d^2\vec{x}\, \rho_{\vec{x}_c}(\vec{x}) r^m e^{im\theta}$$

$$= Ne \int d^2\tilde{x}\, \rho_0(\tilde{x})(r_c e^{i\theta_c} + \tilde{r}e^{i\tilde{\theta}})^m$$

$$= \frac{Ne}{\pi\xi(\sigma_x^2+\sigma_y^2)}\sum_{n=0}^{m} C_m^n (r_c e^{i\theta_c})^{m-n} \int_0^{2\pi} d\theta\, e^{in\theta} \int_0^\infty dr\, r^{n+1} e^{-\frac{r^2}{\xi^2(\sigma_x^2+\sigma_y^2)}}[1-\cdots]$$

$$= Ne \sum_{n=0}^{[\frac{m}{2}]} (2n-1)!! C_m^{2n} (x_c + i y_c)^{m-2n} (\sigma_x^2 - \sigma_y^2 + i\, 2\rho\sigma_x\sigma_y)^n \qquad (19)$$

where $\theta_0 \equiv \arctan \frac{\sigma_x^2-\sigma_y^2}{2\rho\sigma_x\sigma_y}$. See the appendix for the integration and notes on the properties of the m-th moment.

In nonlinear map measurements, we are interested in large off-axis motion, where $x_c \gg \sigma_x$ and $y_c \gg \sigma_y$. So, taking the $n = 0$ terms in Eq.(19) and combining Eqs.(14-17) yields a simple estimates of energy and momentum kicks on beam centroid in one turn:

$$\Delta\delta \sim -\frac{Nr_e}{\gamma} k_l^{(0)} \left[1 + 2 \sum_{m=1}^\infty (\frac{r_c}{b})^{2m}\right] = -\frac{Nr_e}{\gamma} k_l^{(0)} \frac{1+(\frac{r_c}{b})^2}{1-(\frac{r_c}{b})^2} \qquad (20)$$

and

$$\Delta\vec{p}_\perp \sim \frac{Nr_e}{\gamma} k_\perp^{(1)} (x_c \hat{x} + y_c \hat{y}) \sum_{m=1}^\infty m(\frac{r_c}{b})^{2m-2}$$

$$= \frac{Nr_e}{\gamma} k_\perp^{(1)} (x_c \hat{x} + y_c \hat{y}) \left[1 - (\frac{r_c}{b})^2\right]^{-2} \qquad (21)$$

where r_e is electron's classical radius, γ is the beam energy. $k_l^{(0)}$ and $k_\perp^{(1)}$ are the loss factor and kick factor for the lowest multipole modes. The nonlinear dependency on the position of beam centroid is plotted in Figure 1.

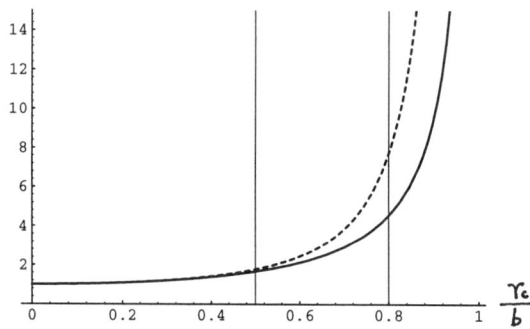

FIGURE 1. Nonlinear factors in the longitudinal (dashed) and transverse wake kicks

Note that, high order contributions are dramatically enhanced when the beam is close to the pipe. A beam less than half way to the pipe wall (i.e.

$r_c \leq 0.5b$) is a comfortable range, and $0.8b$ should be the upper limit. Within this range, the nonlinear factors in $\Delta\delta$ and $\Delta\vec{p}_\perp$ are roughly the same.

Eq.(21) can be viewed as a thin-kick map. Normalizing the dynamical variables to 10σ, we get various order map coefficients. They are no larger than the linear coefficient, which reads

$$k_w \equiv \frac{Nr_e}{\gamma} k_\perp^{(1)} \left(\frac{10\sigma_x}{10\sigma_p}\right) \simeq \frac{Nr_e}{\gamma} k_\perp^{(1)} \bar{\beta} \qquad (22)$$

where $\bar{\beta}$ is the average beta value of the ring. The dimensionless k_w provides a convenient way to estimate the significance of wake effect if one knows the kick factor from machine design or other experiments. By comparing k_w with the accuracy of map coefficient measurement, one can adjust the trade-off between BPM resolution and wakefield effects. Usually, a value less than 10^{-3} is negligible.

Take PEP-II low energy ring as an example [9], where $k_l^{(0)} \simeq 10V/pC$, $k_\perp^{(1)} \simeq 200V/pC\,m$, $\bar{\beta} \simeq 10m$, $\gamma = 6000$. For a $N = 10^{10}$ particle bunch, we have $k_w \simeq 10^{-3}$. Therefore, the transverse wakefield effects is probably negligible. Longitudinally, $\frac{Nr_e}{\gamma} k_l^{(0)} \simeq 5 \times 10^{-6}$ is much less than required energy resolution also. For the high energy ring, the impedances are about the same while energy is 3 times higher and $\bar{\beta}$ is about a factor of 2 smaller. Thus the wakefield effects on PEP-II map measurements should not be a concern. For old machines where impedances have not been carefully budgeted, wakefield could be a limitation on map measurements.

ACKNOWLEDGEMENT

The author C. X. Wang would like to thank A. Chao, G. V. Stupakov, and S. Heifets for their help. This work is supported by the Department of Energy under contract DE-AC03-76SF00515 and the Applied Physics Department of Stanford University.

REFERENCES

1. C. X. Wang and J. Irwin, Possibility to Measure the Poincaré Section Map of a Circular Accelerator. (to be published)
2. R. E. Meller et al., SSC report SSC-N360(1987)
3. I. C. HSU, Particle Accelerator, 1990, v.34, p.43
4. H. Moshammer, Phys. Rev. E 48 (1993) 2140.
5. M. G. Minty et al., Proceedings of the 1995 Particle Accelerator Conference, p.3037 (1995)
6. G. V. Stupakov and A. W. Chao, Proceedings of the 1995 Particle Accelerator Conference, p.3288 (1995)

7. PEP-II An Asymmetric B Factory Conceptual Design Report, SLAC-418
8. A. W. Chao, Physics of Collective Beam Instabilities in High Energy Accelerators, John Wiley & Sons Inc. 1993
9. S. Heifets et al., Impedance study for the PEP-II B-factory, SLAC/AP-99 (1995)

APPENDIX

Before doing the integration, notice that the angle integral has the property

$$\int_0^{2\pi} d\theta e^{in\theta + b\sin(2\theta+\theta_0)} = \int_{-\pi}^{\pi} d\theta e^{in\theta + b\sin(2\theta+\theta_0)} = e^{in\pi} \int_{-\pi}^{\pi} d\theta e^{in\theta + b\sin(2\theta+\theta_0)}$$

which means n must be even.

The double integral can be done as follows and has a closed form.

$$\int_0^\infty dr\, r^{n+1} e^{-\frac{r^2}{\xi^2(\sigma_x^2+\sigma_y^2)}} \int_0^{2\pi} d\theta\, e^{in\theta + \frac{r^2\sqrt{1-\xi^2}}{\xi^2(\sigma_x^2+\sigma_y^2)}\sin(2\theta+\theta_0)}$$

$$= e^{-in\theta_0/2} \int_0^\infty dr\, r^{n+1} e^{-\frac{r^2}{\xi^2(\sigma_x^2+\sigma_y^2)}} \int_{-\pi}^{\pi} d\theta\, e^{i\frac{n}{2}\theta + \frac{r^2\sqrt{1-\xi^2}}{\xi^2(\sigma_x^2+\sigma_y^2)}\sin\theta}$$

$$= 2\pi e^{-in\theta_0/2} \int_0^\infty dr\, r^{n+1} e^{-\frac{r^2}{\xi^2(\sigma_x^2+\sigma_y^2)}} J_{\frac{n}{2}}\left(i\frac{r^2\sqrt{1-\xi^2}}{\xi^2(\sigma_x^2+\sigma_y^2)}\right)$$

$$= \pi e^{-in\theta_0/2}[\xi^2(\sigma_x^2+\sigma_y^2)]^{\frac{n}{2}+1} \int_0^\infty dr\, r^{\frac{n}{2}} e^{-r} J_{\frac{n}{2}}(i\sqrt{1-\xi^2}\,r)$$

$$= \pi e^{-in\theta_0/2}[\xi^2(\sigma_x^2+\sigma_y^2)]^{\frac{n}{2}+1} \frac{(2i\sqrt{1-\xi^2})^{\frac{n}{2}} \Gamma(\frac{n}{2}+\frac{1}{2})}{\sqrt{\pi}(\xi^2)^{\frac{n}{2}+\frac{1}{2}}}$$

$$= \pi\xi(\sigma_x^2+\sigma_y^2)[i\, e^{-i\theta_0}\sqrt{1-\xi^2}(\sigma_x^2+\sigma_y^2)]^{\frac{n}{2}} \frac{\Gamma(\frac{n+1}{2})2^{\frac{n}{2}}}{\sqrt{\pi}}$$

$$= \pi\xi(\sigma_x^2+\sigma_y^2)(\sigma_x^2-\sigma_y^2 + i\,2\rho\sigma_x\sigma_y)^{\frac{n}{2}}(n-1)!!$$

Notes on the m-th moment expression:

- The real (imaginary) part gives the normal (skew) component.

- Exact to all orders for Gaussian distribution, and to $m \leq 2$ (quadrupole mode) for any distribution.

- For on-axis beam ($x_c = y_c = 0$), there are only even-order moments. Thus no transverse kick to the bunch centroid as expected from symmetry consideration.

- For large off-axis beam, the first term ($n = 0$) dominates, i.e. the bunch behaves as a marco-particle.

Application of Moments Method to Dynamics of Muon Cooling System

Zohreh Parsa[†1] and Pavel Zenkevich[‡]

[†] *Department of Physics, Bldg. 901A*
Brookhaven National Laboratory
PO Box 5000 Upton, NY 11973-5000

[‡] *Institute for Theoretical and Experimental Physics*
B Cheremushkinskaya Ulitsa 25, RU-117 259, Moscow, Russia

Abstract. Application of moments method to study dynamics of muon cooling system is presented. The muon cooling channel has specific features such as damping due to ionization cooling, heating due to multiple scattering and energy struggling. Our method (in a case of transverse motion) is correct in absence of longitudinal magnetic fields such as solenoids.

INTRODUCTION

It is well known that moments method is successfully used for calculation of the dynamics in different focusing channels. However, the muon cooling channel has some specific features: a) damping due to ionization cooling; b) heating due to multiple Coulomb scattering and energy struggling. In our paper [1] on "Kinetics of Muon Longitudinal Cooling" we considered an evolution of longitudinal beam moments in a system with constant parameters. In this paper we investigate a more general case: a cooling system, consisting of a number of sections, each of which has constant parameters. Our method (in a case of transverse motion) is correct in absence of longitudinal magnetic field (e.g. solenoids).

---

[1]) Supported by US Department of Energy Contract No. DE-AC02-76CH00016 and National Science Foundation NSF-PHY-94-07194

METHOD OF MOMENTS

Equation of motion for individual particle in presence of damping and random heating may be written as:

$$y'' + 2\alpha y' + w^2 y = W(z) \tag{1}$$

The y is a particle coordinate (transverse or longitudinal one), 2α is a damping coefficient, w is a gradient of the external field, $W(z)$ is a function describing heating.

We can consider $W(z)$ as a sum of random δ-functions (kicks), with given spectral density:

$$\frac{d\langle y'^2 \rangle}{dz} = W \tag{2}$$

with $y' = dy/dz$. Let us assume that W is also constant. Then, by simple differentiation we find (using Eq. 1), that

$$\left. \begin{array}{rcl} \frac{d\langle y^2 \rangle}{dz} & = & \langle yy' \rangle \\ \frac{d\langle yy' \rangle}{dz} & = & \langle (y')^2 \rangle - 2\alpha \langle yy' \rangle - \Omega^2 \langle y^2 \rangle \\ \frac{d}{dz}\langle (y')^2 \rangle & = & 4\alpha \langle (y')^2 \rangle + 2\Omega^2 \langle yy' \rangle + W \end{array} \right\} \tag{3}$$

Let $\langle y^2 \rangle = u$; $\langle yy' \rangle = v$; $\langle (y')^2 \rangle = t$. We can rewrite our system in the following form

$$\left. \begin{array}{l} u' - 2v = 0 \\ \Omega^2 u + v' + 2\alpha v - t = 0 \\ -2\Omega^2 v + t' + 4\alpha t = W \end{array} \right\} \tag{4}$$

This system should be solved for general initial conditions:

$$u = u_0; \quad v = v_0; \quad t = t_0. \tag{5}$$

The solution of the linear system (4) can be written in matrix form:

$$\begin{pmatrix} u_f \\ v_f \\ t_f \end{pmatrix} = W \begin{pmatrix} R \end{pmatrix} + \begin{pmatrix} M \end{pmatrix} \begin{pmatrix} u_0 \\ v_0 \\ t_0 \end{pmatrix} \tag{6}$$

Here subscript "f" corresponds to final point of the section, (R) is a column and (M) is 3×3 matrix.

Elements of R and M may be found using Laplace transform method. We have the following steps:

1. Using Laplace transform method we get a system of linear equations for $U(p); V(p); T(p) \rightarrow$ Laplace transforms of $u(z); v(z); t(z);$ (p is a Laplace variable).

2. Solving this linear system we obtain expressions for $U(p); V(p); T(p)$.

3. A backward Laplace transform can be found using a residue theorem:

$$f(z) = \sum_k \text{res } F(p_k) \exp(p_k z') \qquad (7)$$

Here summation is made over all p_k which are the first order residues of function $F(p)$.

Using this procedure, we can obtain final expressions for all the elements.

$$M_{1,1} = \exp(-2\alpha z)[ch2\alpha_0 z + \frac{\alpha}{\alpha_0} sh2\alpha_0 z +$$
$$\frac{\Omega^2}{2\alpha_0^2}(ch2\alpha_0 z - 1)] \qquad (8)$$

$$M_{1,2} = \exp(-2\alpha z)\left[\frac{\alpha}{\alpha_0^2}(ch2\alpha_0 z - 1) + \frac{sh2\alpha_0 z}{\alpha_0}\right] \qquad (9)$$

$$M_{1,3} = \exp(-2\alpha z)\frac{ch2\alpha_0 z - 1}{2\alpha_0^2} \qquad (10)$$

$$M_{2,1} = \exp(-2\alpha z)\left[\frac{\Omega^2 \alpha}{2\alpha_0^2}(1 - ch2\alpha_0 z) - \frac{\Omega^2}{2\alpha_0}sh2\alpha_0 z\right] \qquad (11)$$

$$M_{2,2} = \exp(-2\alpha z)[\alpha(ch2\alpha_0 z - 1) + \alpha_0 sh2\alpha_0 z] \qquad (12)$$

$$M_{2,3} = \exp(-2\alpha z)\left[\frac{\alpha}{2\alpha_0^2}(1 - ch2\alpha_0 z) + \frac{sh2\alpha_0 z}{2\alpha_0}\right] \qquad (13)$$

$$M_{3,1} = \exp(-2\alpha z)\left[\frac{\Omega^4}{2\alpha_0^2}(ch2\alpha_0 z - 1)\right] \qquad (14)$$

$$M_{3,2} = \exp(-2\alpha z)\Omega^2 \left[\frac{\alpha}{\alpha_0^2}(1 - ch2\alpha_0 z) + \frac{sh2\alpha_0 z}{\alpha_0}\right] \qquad (15)$$

$$M_{3,3} = \exp(-2\alpha z)\left[\frac{\Omega^2}{2\alpha_0^2}(ch2\alpha_0 z - 1) + (ch2\alpha_0 z - \frac{\alpha}{\alpha_0}sh2\alpha_0 z)\right] \qquad (16)$$

All diagonal terms ($M_{1,1}$, $M_{2,2}$, and $M_{3,3}$) should be equal to 1 for $z = 0$, all others are equal to 0 for $z = 0$. For column:

$$R_1 = \frac{1}{4\alpha\Omega^2}[1 + \frac{\Omega^2}{\alpha_0^2}\exp(-2\alpha z) - \frac{\alpha\exp(-2\alpha z)}{\alpha_0^2}$$

$$(\alpha_0 sh2\alpha_0 z + \alpha ch2\alpha_0 z)] \tag{17}$$

$$R_2 = \frac{\exp(-\alpha z)}{4\alpha_0^2}(ch2\alpha_0 z - 1) \tag{18}$$

$$R_3 = \frac{1}{4\alpha\Omega}\left\{1 + \exp(-2\alpha z)\left[sh2\alpha_0\left(1 - \frac{\alpha}{\alpha_0}\right) - \frac{\alpha^2 ch2\alpha_0 z - \Omega^2}{\alpha_0^2}\right]\right\} \tag{19}$$

Here

$$\alpha_0 = \sqrt{\alpha^2 - \Omega^2} \tag{20}$$

If α_0 is imaginary, then

$$\left.\begin{array}{c} ch\alpha_0 z \to \cos w_1 z \\ \frac{sh\alpha_0 z}{\alpha_0} \to \frac{\sin w_1 z}{w_1} \end{array}\right\} \quad w_1 = \sqrt{\Omega^2 - \alpha^2} \tag{21}$$

APPLICATIONS TO LONGITUDINAL COOLING

In this case $y = z - z_s$. (z_s is a coordinate of equilibrium particle),

$$2\alpha = \frac{1}{mc^2\beta^2\gamma}\left(-\frac{dE}{dz}\right)_{ion}\frac{\Psi}{L_0} + \frac{d}{dz}\left(\frac{p}{\Lambda}\right)/\frac{p}{\Lambda} - \frac{d}{dp}\left[\left(\frac{dE}{dz}\right)_{ion}\frac{1}{v}\right] = 2\alpha_1 + 2\alpha_2 + 2\alpha_3 \tag{22}$$

The first term in RHS describes longitudinal cooling due to wedges, the second one damping due to change of longitudinal mass, and the third one — so named "natural" cooling due to slope of the ionization losses curve [2]. Here

$$L_0 = \frac{\left(\frac{dE}{dz}\right)_{ion}}{\frac{d}{dX}\left(\frac{dE}{dz}\right)_{ion}} \tag{23}$$

We see, that if $\left(\frac{dE}{dz}\right)_{ion}$ does not depend on X, the derivative is equal to zero, and the first term in (22) disappears. Where X is a transverse coordinate.

Parameter Λ is defined by

$$\Lambda = \left\langle\frac{\Psi}{R}\right\rangle + \frac{1}{\gamma^2} \tag{24}$$

R is a radius of curvature for ideal trajectory. Ω is the synchrotron frequency and in linear approximation may be written as

$$\Omega^2 = \frac{2\pi E_{ac}\cos\left(\frac{2\pi}{\lambda}z_s\right)\Lambda}{mc^2\lambda\beta^2\gamma} \tag{25}$$

Here E_{ac} is an amplitude of r.f. field, and λ is its wavelength.

$$W_y = \left(\frac{p}{\Lambda}\right)^{-2} \frac{1}{\beta^2 mc^2} \frac{d}{dz}\langle(\Delta E)^2\rangle \tag{26}$$

$$\frac{d}{dz}\langle(\Delta E)^2\rangle = K_s\gamma^2 \left(1 - \frac{\beta^2}{2}\right) \tag{27}$$

$$K_s = 4\pi(r_e m_e c^2)^2 \frac{\rho Z}{A} \tag{28}$$

Here r_e, m_e — classical radius and mass of electron; ρ; Z; A are density; charge and atomic number of the wedge material m - muon mass. Let us consider sections which are specific for cooling system:

a) <u>Wedge section</u>
$\alpha = \alpha_1 + \alpha_2 + \alpha_3$; $W = W_y$; $\Omega_s^2 = 0$

b) <u>Section of transverse cooling ($\Psi = 0$)</u>
$\alpha = \alpha_2 + \alpha_3$; $W + W_y$; $\Omega_s^2 = 0$

c) <u>Accelerating section</u>
$\alpha = \alpha_2$; $W = 0$; $\Omega_s^2 \neq 0$

APPLICATIONS TO TRANSVERSE COOLING

Here y is a transverse coordinate

$$\alpha = -\left(\frac{dE}{dz}\right)_{ion} \frac{1}{mc^2\beta^2\gamma} - K\alpha_1 + \frac{p'}{p} \tag{29}$$

K is a coefficient describing coupling with "wedge section", if $\Psi = \Psi_y$, $K = 1$; if $\Psi_y = 0$, $K = 0$.

$$W = W_\perp = \frac{E_s^2}{(pc\beta)^2} \frac{1}{X_r} \tag{30}$$

$E_s = 15$ MeV, X_r is material radiation length. Let us consider sections which are specific for ionization cooling system.

a) <u>Wedge section</u>
$W = W_\perp$
$\alpha = \alpha_\perp^0 - K\alpha_1 + \frac{p'}{p}$
It is easy to show, that $\frac{p'}{p} + \alpha_\perp^0 = 0$,
Ω^2 is arbitrary

b) <u>Section of transverse cooling</u>
   $W = W_\perp$
   $\alpha = 0$
   Ω^2 is arbitrary

c) <u>Accelerating section</u>
   $W = 0$
   $\alpha = \frac{p'}{p}$
   Ω^2 is usually zero.

Since in our variables X, $\frac{dX}{dz}$; damping rate (in absence of wedges) is equal to zero, if wedges are used, then we have increment ($\alpha < 0$) in such a section.

ACKNOWLEDGEMENTS

This work was partially carried out at the Institute for Theoretical Physics, University of California, Santa Barbara, as part of the New Ideas for Particle Accelerators Program with partial support by the the National Science Foundation Grant No. NSF-PHY-94-07194.

REFERENCES

1. Z. Parsa, P. Zenkovich, Kinetics of Muon Longitudinal Cooling, to be published (1997).
2. V. Parchomchuk and A. Skrinsky, *AIP CP* **372**, 139 (1996).

On 2D Electron Cloud Dynamics In High-Current Plasma Lens For Ion Beam Focusing.

A. A. Goncharov, I. V. Litovko, I. N. Onishchenko, V. F. Zadorozhny *

Institute of Physics National Academy of Science of Ukraine, Prospect Nauki 46, Kiev 28, 252450, Ukraine.
* Institute of Cybernetic National Academy of Science of Ukraine, Prospect Glushkova 20, Kiev, 252187

Abstract. In this paper we are dealing with the appear the stable existence and dynamics of 2-D electron vortical structures in crossed electric and magnetic fields. The collective interactions in which the electron motion is nonlinear and ion motion is linear, is concerned. By using of the kinetic equation and the catastrophe theory approach we deduce an origin of the vortical structures. The nonlinear differential equation for the electric potential in a hydrodynamical approximation is obtained. It describes a drift motion of the electrons in oscillating electic fields of the high-current plasma lens(PL), arising due to presence the principal unremoval radical gradient of the axical component of the magnetic field. It was shown that the considered equations have contained the solutions in the form of the single vortical structures. The stability of the structures are given.

INTRODUCTION

Localized structure or soliton-like solution arises by means of a sufficiently large influence of the outside forces and also of the nonlinear effect of the self influence. There are the unusually properties (for linear dynamics) for objects which are formed in this case, the stability is one of them. The study those localized structure is the topic of the physic of the soliton.

By means of the localized structure we can eliminate the discrepancy into singular points. That is why these causes Lord Kelvin had offered to consider the "vortical atom" instead of the point atoms. Similar sentence were express one's views by O. Hevisaid, J. J. Tompson, T. Mi and so on. A. Einstein has studied this problem more concret. The assumption was made by him that the particle may be described by means of the regular solutions of some nonlinear equations. This solutions have to be some bunched field occupying the bounded region in the space and this region a tension of the field or the energy of a density became very large [1]. After almost 100 years of the break it taking to study of the quantum world, physicists are busy studying of the vortical structures as the soliton-like solutions. Outstanding contribution in this topic made Vlasov [2].

The story the structures localized in a finite region in a plasma have less then 20 years. Nonlinear equations the corresponding potential drift wave in plasma was studying in [3]. Here has proved the formation of 2-D soliton as result a drift unstable of the magnetic plasma. A soliton structure (excellent described by Alan C. Newell [4] without the attention don't remained of the magnetic active plasma in which act the Lorentz force. Gradually a clear became that the presence of the vector nonlinear which can be regarded as a special Jacobian for equations of the 2-D vortex structure yield a new quality for the connections which have not KdV [5].

A like with the magnetic confined traps exist a wide class of the systems with a plasma in magnetic field. Particulary the plasma optic systems with magnetic isolation of cold electron and free nonmagnetized ions [6]. Particulary, one special there are drift unstable by a gradient isolating of magnetic field. In [7] has considered the influence of the gradient of a isolating magnetic field applied to PL on the arising collective drift ion-electron instability. Nevertheless, there are the question of the exist of global solutions its evolution and stable still remain bottle-neck. One determine main difficulty for the analysis of the complex motion. Here the principal topic for this aim the Lyapunov's direct method are discussed. For the arise of the vortex we focus mainly on the theory of the non-Morse points, which possibilities in the description of structures localized in a finite region and one connection with global solution of the Vlasov's equation and its approximate gydrodynamical solutions [8].

VLASOV'S EQUATION AND ITS RESOLUTION

It will be consider the dynamic 2-D electron cloud of the PL. As compared to the electrons the ions have been unmobile.

It should be considered here the mathematical basic of this problem is the Vlasov-Poisson equation, as well known. We shall try to present these question in a different way and to precise some new aspects concerning to the existence of some potential in phase space. The questions of the existence and stability of the solutions are the main for this problem. And we try not to move away of these questions. Thus, we have to study the problem arising of the vortex in electron cloud more precise and vigorous.

Now there are many problems of plasma dynamics which are resolved without the theorems of the existence of the solution and without the estimations of the approximations. Hereby we are interesting in the research of general solution for the research of general solution for the following equation:

$$\frac{\partial f}{\partial t} + \nabla_x f \cdot v + \nabla_v f \cdot <\dot{v}> = 0 \qquad (1)$$

It's well known Vlasov's equation for distributive function $f(t,x,v)$. We will be examine here 2D problem only. Consequently, on this cause $\vec{x} = (x, y), \ldots \vec{v} = (v_x, v_y)$ where x- is vector of the position, v- is vector of the velocity. Here

$$<\dot{v}> = \frac{\int \dot{v} f(t,x,v,\dot{v}) d\dot{v}}{f(t,x,v)} = \frac{e}{m}\left(\vec{E} + \frac{1}{c}[\vec{v} \times \vec{H}]\right) \qquad (2)$$

c - is charge of the electron, m - it's mass. H - is magnetic field. For electrostatical potential φ is Poisson equation:

$$\Delta \varphi = -4\pi e \int f dv \qquad (3)$$

We assume that H - is parameter which we given. Notes here, that $\rho = \int f\, dv$ is a density of electron charge. A rigorous result analysis must be based at the theorem of the existance of solution of the (1), (3). For this aim we will look for the solutions at the form:

$$f = f_0(x,v) e^{i\omega t}$$

Consequently
$$L_0 f = -i\omega f$$

We denote by L_0 operator

$$L_0 f = <\nabla_x f \cdot v> + <\nabla_v f \cdot <\dot{v}>>$$

where $<x,y>$ is scalar product in Euclid space.

Now we will be prove that the system (1), (3) has been integrated. It may be noted that the dynamical system which generator of the quasi-linear operator L_0 will be

$$\frac{dx}{dt} = v \qquad (4)$$

$$\frac{dv}{dt} = \frac{e}{m}\nabla\varphi + \frac{1}{c}[v \times H]$$

If the system (4) expand introducing a new coordinate f_0:

$$\frac{df_0}{dt} = -i\omega f_0 \qquad (5)$$

then the system (4), (5) will be a linear system. One save some measure so as the following condition $I_t = I_0$ holds for all t. Here I_t is phase volume for a time $t \geq 0$. By theorema Gelgand - Kostuchenko [ ] and system (4), (5) we conclude that equation (3) will be resolved if following condition holds:

$$i\omega \in \sigma(L_0)$$

where $\sigma(L_0)$ - is spectrum of the operator L_0. Let us denote a vector $\overset{\Delta}{X} = (v, <\dot{v}>)$. Since the vector field X fulfilled condition $div\ X = 0$ the potential V must to be that the following relation holds $\nabla V = X$, consequently $\Delta V = 0$, $\Delta_x V$ is identicaly equal to zero. Thus:

$$\frac{d^2V}{dv_1^2} + \frac{d^2V}{dv_2^2} = 0 \qquad (6)$$

We study the problem (6) in a bounded, regular and connected two dimensional domain $\overline{\Omega} = \{v_1, v_2 : v_1^2 + v_2^2 \leq v_0^2\}$ with several types of boundary conditions are concidered.

The first of one is the Dirichlet boundary condition corresponding to Ω is circle function V have to be harmonic and there is condition on the boundary $\partial\Omega$: $V = \Psi(v)$, $v_1, v_2 \in \partial\Omega$; here a function $\Psi(v)$ is proposed and one is continued and periodical. Let the center of circle Ω will be the initial coordinates. In particular $0 \in z$, i.e. a initial have to lay on the axis z. We introduce next the polar coordinates $\{|v|, \theta\}$, here: $v_x = |v|\cos\theta$, $v_y = |v|\sin\theta$ and $V = V(|v|, \theta)$.
Now according to (6) we write down:

$$\frac{\partial}{\partial|v|}\left(|v|\frac{\partial V}{\partial|v|}\right) + \frac{1}{|v|}\frac{\partial^2 V}{\partial\theta^2} = 0 \qquad (7)$$

Let $V = \eta(\theta)\cdot\Lambda(|v|)$ then

$$= \frac{A_0}{2} + \sum_{n=1}^{\infty}\left(A_n\cos(n\theta) + B_n\sin(n\theta)\right)\cdot|v|^n \qquad (8)$$

$$(|v|, \theta)_{|v|=|v_0|} = \Psi(\theta)$$

and $A_n = \int_0^{2\pi} \Psi(\theta)\cos(n\theta)d\theta$, $B_n = \int_0^{2\pi} \Psi(\theta)\sin(n\theta)d\theta$. $A_0 = 0$ for periodical function. Now our task consist in the finding of the general solution of the equation (3) at the terminology potential V. We will be to find the function f for which following equal holds: $f = f(V)$. By this assumption and according to (3) may be obtain the following solution:

$$f_0 = f_{00} e^{i\frac{\omega}{D}V}$$

where $(\nabla V)^2 = D$. It is easy to see that $D = \text{const}$ f_{00} initial value of the f_0.
Now

$$f = f_{00} e^{-i\omega t} \exp\left(ik\left(\frac{A_0}{2} + \sum_{n=0}^{\infty}(A_n \cos(n\theta) + B_n \sin(n\theta))|v|^n\right)\right) \quad (9)$$

or:

$$f = f_{00} e^{-i\omega t} \exp\left(\frac{ik}{2\pi}\int_0^{2\pi} \Psi(\xi)\frac{(v_0^2 - v^2)d\xi}{v_0^2 - 2vv_0\cos(\xi - \theta) + v^2}\right)$$

For the first approach we have

$$f(t, x, v) = f_{00} \exp i(k(\alpha x_1 + \beta x_2) - \omega t)$$

where α, β some constants.

It's some oscilating process as main process and the remaining portion of the series of the Fourier has been take birth to the ruffly waves. By means of the variations of the boundary functions $\Psi(\theta)$ we may study different of the waves processes. But we omit this step. We describe now other properties of potential V. Suppose that V is multivalued. Denote this position may be when a function $\Psi(\theta)$ is multivalued or explosive.

It is easy to see that following equation holds:

$$= \left(Ae^{l\theta} + Be^{l\theta}\right)\left(C|v_0|^l + D|v_0|^{-l}\right) \quad (10)$$

here A, B, C, D some constants, l - mode.

ARISING VORTEX

In this item we are dealing with singular points of the function $V(|v|, \theta)$ ang we give sufficient conditions concerning the arising and stability of singular vortex.

Definition 1. The points for which the equality $\nabla V = 0$ holds, we named of the Morse's critical points of the function of 2 variables $V(|v|, \theta)$, if $\det|V_{ij}| \neq 0$ [i,j =1,2].

Definition 2. The critical points of the function $V(|v|,\theta)$ for which $\det|V_{ij}| = 0$ we named the non Morse's singular points.

Definition 3. The least power of a term of the Taylor series in the point $p(|v|, \theta)$ will be called a off-spring of the function $V(|v|, \theta)$ in point p.

Now consider the existence the Morse and non-Morse singular points and its off-spring contribution an appearence of a vortex. It's easy to see that any point $p_0(|v|, \theta)$ for which $\nabla \varphi = -[v, H]$ will be the Morse's critical point. Thus the Lorentz force in this point equal to zero, i.e. in this point p_0 obey of the equilibrium.

We have arrived at the following result: the critical Morse's points are not the vortex points. This conclusionb based on the introduced concept of simply potential V to the phase velocity, as well known.

Now suppose that the system (1) possesses the non-Morse singular point $p_1(|v|, \theta)$, i.e.

$$\det|V_{ij}| = \eta_{\theta\theta} \Lambda_{|v||v|} \eta \Lambda - \eta_\theta^2 \cdot \Lambda_{|v|}^2 = 0$$

Theorem 1. The non-Morse singular point p_1 will be at the system (1) if and only if the potential V has been multivalued function.

Proof. Above we establish that the simply potential can't bring about or beget of the vortex motion. Hereby so as to one will be the boundary function $\Psi(\theta)$ necessarily must be of the multivalued function. A sufficient of the existence of the vortex in these conditions has followed from the existence the closed trajectory in any small neighbourhood. This statment has from the condition $\det |V_{ij}| = 0$ consequently, we can construct a function Lyapunov in some neughbourhood of the pont p_1 which satisfy of the theorem Lyapunov about stable. From this theorem follow the fact of the existence of the simple closed lines.

Remark 1. For PL the function $\Psi(\theta) = [v \times H]|_{v=v_0}$. One is multivalued function becouse of the velocity is multivalued. We have arrived at the following result: Magnetic field has begeted single vortex.

Remark 2. We have presented conection between *rot* V and *rot* u, where u is hydrodynamical velosity of an electron. In order to get the *rot* u we will write $u = \int v \cdot f_0 \cdot dv$, coming back to the equal $f_0 = f_{00} e^{-V/D}$ and using of the operator *rot* we get the identity:

$$rot(u) \hat{=} \int vf \cdot \frac{1}{D} rotV dv$$

because of the velocity v is the phase coordinate and is don't depended of vector x. Consequently, in according to the definition *rot u* mentioned above one vanishes together with disappearance of the vortex V.

Remark 3. It's known: any vector can be presented as sum $v = rot\ A + grad\ b$, becouse of velocity may be rewritten as sum multivalued function and symply function. Thus the electron cloud consist from waves motions and vorticity.

Remark 4. It the potential V in point p_1 has local minimum then vortex motion will be stable. This conclude is following from the exist Lyapunov's function: $S(y) = V(p_1) - V(p_1+y)$, where $y = p - p_1$ in some neighbourhood of the point p_1, $S(0) = 0$; $S(y) > 0$, and

$$\frac{dS}{dt} = -(\nabla V)^2 \leq 0$$

So then the conditions of the theorem Lyapunov about stability is fulfilled. So that a solution $y=0$ is stable, i.e. vortex in point p_1 is stable. The existence of spring-off is a new result for this problem. One stability and unstability give birth to the new structure of the dynamical system. For example one point with det $|V_{ij}| = 0$ give two points with $\nabla V = 0$ and det $|V_{ij}| \neq 0$ i.e. a monopolar vortex make decoupling on two vortex. Exterior disturb monopolar vortex introduced in depolar vortex by jump. Thus disturbance of the function V in non-Morse point take birth to a new qualitative of the dynamical structs [8].

Remark 5. Let $A=B$, $N=A \cdot C$ and $D_n=0$ then $V=N(e^{k(\theta-\theta_0)}+e^{-k(\theta-\theta_0)})|v_0|=Nch(k(\theta-\theta_0))\cdot|v_0|$ or:

$$f = \frac{f_{00}}{e^{i\omega t} \exp(Nch(k(\theta-\theta_0))|v_0|)} \qquad (11)$$

Thus there is Kadomtsev - Petviashvili soliton-like solution with two parameters ($|v_0|, \theta_0$).

SOME USEFUL APPROACH AND IT STABILITI.

In this item we discuss the some nonlinear approach which have interesting soliton - like solutions. Let us consider the transformation of the equation of the Vlasov on hydrodynamic equation Navier - Stokes. For this aim let us multiply equation (1) by an arbiotrary function of velocity $v-\alpha(v)$. Let

$$< \dot{v} > f \big|_{-\infty}^{+\infty} \to 0$$

sufficient fast so that

$$\frac{\partial}{\partial t}\int \alpha f dv + div_x \int v \alpha f dv = \int \frac{\partial \alpha}{\partial v_s} <\dot{v}>_s f dv$$

holds. In hydrodynamical model in cold electron approximation, assume that $\omega/\omega_{he} \ll 1$ (drift approximation), where $\omega_{he} = eH/mc$- cyclotron frequency, in first approximation we have $v = \frac{e}{m\omega_{he}}[\vec{e}_z \times \nabla\varphi]$. Here $H = H(x)$ - magnetic field. Assume that the $\alpha(v) = 1$, s=1,2 and performed the opertations integration by parts and div_x we have the following equation in this approximation :

$$\frac{\partial \Delta \varphi}{\partial t} - \frac{e}{m}\nabla_y \varphi \Delta \varphi \frac{\partial}{\partial x}\left(\frac{1}{\omega_{he}}\right) + \frac{e}{m\omega_{he}}\{\varphi, \Delta\varphi\} = 0 \qquad (12)$$

where $\{a,b\} = \frac{\partial a}{\partial x}\frac{\partial b}{\partial y} - \frac{\partial a}{\partial y}\frac{\partial b}{\partial x}$. Equation (12) conserved integral of energy and potential vorticity $\Omega = rot\, v + \omega_{he}$, because it conserved symmetry of hydrodynamical equations. In linearing approximation give from (12) dispersion equation for small oscillation [7]:

$$\chi = \frac{k_y \frac{\partial}{\partial x}\left(\frac{n_0}{\omega_{he}}\right)}{k^2\left(\omega - k_y \frac{eE_0}{m\omega_{he}}\right)} \qquad (13)$$

where E_0 - stationary electrical field PL, $E_0=E(k)$, n_0 - stationary density of electrons. Equation (13) describe drift instability in strongprecise behavior, even if stationary density n_0 is homogeneous in section, if value of magnetic field grows along radius. In weakprecise behavior, when may be neglect of fast ions of beam, equation (13) will be describe of the diocotron unstability dissimilar electronical cloud .

We will search stationary solution (12) which moving with velocity u along y-axes in form $\varphi=\varphi(x, y-ut)$. Then we obtain:

$$\left\{\varphi - \frac{u}{c}\int H(x)dx, \frac{\chi\Delta\varphi}{H(x)}\right\} = 0$$
$$(14)$$

Solution of this equation is

$$\frac{\Delta \varphi}{H} = F(\varphi - \frac{u}{c}\int H(x)dx) \qquad (15)$$

Where F is any function. From sort F depend behavior of solution: it can be wave type or localized in some region. We are interesting solution which describe structure localized in region with radius *a* in some point (x_0, y=ut). Then select F such sort:

$$\Delta \varphi = -k^2 \frac{H(x)}{H_0}\left(\varphi - \frac{u}{c}\int H(x)dx\right), \qquad 16)$$

$$r \leq a$$

$$\Delta \varphi = 0, \qquad r \geq a$$

where k is constant, $x=x_0+x$, $r^2=x^2+(y-ut)^2$.
For case when $H(x)=H_0$=const equation (16) have solution analogous to solution of Larichev- Reznik [9] (compare with [10]):

$$\varphi = C_0(J_0(kr) - J_0(ka)) + \frac{u}{c}Hx(1 - \frac{2J_1(kr)}{krJ_0(ka)}) + \frac{u}{c}Hx_0, \qquad (17)$$

$$r \leq a$$

$$\varphi = \frac{u}{c}Hx(\frac{a}{r})^2 + \frac{u}{c}Hx_0, \qquad r \geq a$$

where $x=r\cos\theta$. This solution satisfy conditions continuosly φ and ∇φ on boundary r=a, on boundary φ - ∫ H(x)dx =0 , from this condition also follow dispersion relation $J_1(kr) = 0$. We have 3 free parameters in (17) - size of structure- *a*, velocity- u and amplitude of potential- C_0. We can see that this solution localized in region size a and fall dawn outside as $1/r^2$.

For case H=H(x) we assume that $H(x)=H(x_0+x)=H(x_0)+(\partial H/\partial x)x_0$ and $x << x_0$, than in 0-approximation solution (16) is analogy (17), in next - linear for *x* approximation we obtain:

$$\varphi = \frac{ua}{2ckJ_1(\kappa a)}\left(\frac{\partial H}{\partial x}\right)_{x_0}(J_0(\kappa r)-J_0(\kappa a)) + \frac{2u}{ck}\frac{H(x_0)}{1-\kappa a\frac{\Phi(\kappa r)}{\Phi(\kappa a)}}\left(\frac{\Phi(\kappa r)}{\Phi(\kappa a)} - \frac{J_1(\kappa r)}{J_1(\kappa a)}\right)\frac{x}{r} - \frac{ua}{ckJ_1(\kappa a)}\left(\frac{\partial H}{\partial x}\right)_{x_0}J_2(\kappa r)(2\frac{x}{r^2}-1)+ \qquad (18)$$

$$+\frac{u}{c}\left(H(x_0)x_0 + H(x_0)x + (\frac{\partial H}{\partial x})_{x_0}\frac{x^2}{2}\right) \qquad r \leq a$$

$$\varphi = \frac{u}{c}\left(H(x_0) + \left(\frac{\partial H}{\partial x}\right)_{x_0}\frac{a^2}{4}\right) - \frac{u}{c}\left(\frac{\partial H}{\partial x}\right)_{x_0}\frac{a^2}{4}\left(\frac{a}{r}\right)^2 + \frac{u}{c}H(x_0)\left(\frac{a}{r}\right)^2\left(x + \left(\frac{\partial H}{\partial x}\right)_{x_0} \cdot \frac{1}{H(x_0)}\left(\frac{x}{r}\right)^2\right), \quad r \geq a$$

where $\kappa = k\sqrt{\frac{H(x_0)}{H_0}}$,

$$\Phi_1(\kappa r) = \pi\left(\frac{\partial H}{\partial x}\right)_{x_0}\frac{1}{H(x}\left(\left(\frac{\partial H}{\partial x}\right)_{x_0} r^2 (J_2(\kappa r) Y_1(\kappa r) + J_1^2(\kappa r)) - \sum_{n=0}^{\infty}(-1)^n \frac{(\frac{\kappa r}{2})^{2n+4}\Gamma(2n+2)n}{(2n+3)\Gamma^3(n+2)}\right)$$

This solution satisfy conditions continuosly φ and $\nabla\varphi$ on boundary r=a. We have 2 free parameters in (18) - size of structure and velocity u, taking into account dispersion relation $J_2(\kappa r) = 0$. From solution (18) we can see that it satisfy conditions for vortex structures : it is localized in some region of size a in center on some point (x_0, y_0) and fall dawn outside in accord power law, velocity of structure $u > v_0$, where v_0 is drift velocity $= cE_0/H_0$ and velocity rotation $v_{rot} > u$.

Now, we consider here the stable of the particular solution ρ of the equation (12). Let φ' will be some disturb so that :

$$v_x = -\frac{\partial}{\partial y}(\varphi + \varphi'), \quad v_y = \frac{\partial}{\partial x}(\varphi + \varphi')$$

Consider following functional $\Phi = \frac{1}{2}\iint_{S(t)} \varphi^2 dx\,dy$, where the boundary $S(t)$ consist with liquid particles. Using the known formule:

$$\frac{d}{dt}\iint_{S(t)} f\,dx\,dy = \iint_{S(t)}\left(\frac{\partial f}{\partial t} + div(vf)\right)dx\,dy$$

These relations allowed to obtain $\frac{\partial \Phi}{\partial t} = \iint_{S(t)} \rho J(\varphi, \Delta\varphi)\,dx\,dy$, where $J(\varphi,\Delta\varphi)$ is Jucobian. We now introduce the functional $F = \frac{1}{2}\int_{-\infty}^{\infty}\int \rho^2 dx\,dy$, using (12) we can statment that with respect to (12) the following relation holds: $\frac{\partial F}{\partial t} = \int_{-\infty}^{+\infty}\int \rho J(\varphi, \Delta\varphi)\,dx\,dy$. Now we shall prove that $\iint \rho J(\varphi, \Delta\varphi) \leq 0$, i.e. the F - is the Lyapunov functional and one determine the stability of the particle solution φ at norm: $\|\rho^2\| = (\rho, \rho) = \iint \rho^2\,dx\,dy$. For this aim we assume ξ= y-ut, and using J(φ,Δφ) we will be get: $\int\int_{-\infty}^{+\infty} \rho dx d\xi = (A\rho, \rho)$, where A is a bound linear operator. Thus we have to prove $(A\rho, \rho) \leq 0$. Consider the subsidiary of the linear Cauchy problem:

$$\frac{\partial u}{\partial t} = A u, \qquad u|_{t=t_0} = u_0 \in H_2(R_2)$$

In [8] have the proof that a spectrum σ of the operator A fulfilled a condition Re σ(A) < 0 i.e. $\frac{\partial F}{\partial t} \leq 0$. This result proved the stability of the particular solution φ.

ACKNOWLEDGMENTS

We was supported by the Science & Technology Center of Ukraine (pr. N 298).

REFERENCES

1. Einshtein A., *Collection of scientific works*, vol. 2, Moscow: "Nauka", 1966
2. Vlasov A. A., *Statistical function of distribution*, Moscow: "Nauka", 1966
3. Petviashvili V. N., Pochotelov O. A., *Journal "Physica Plasmy"*, 12, p. 1127 (1986)
4. Newell C. Alan, *Solitons in mathematics and physics*, 1989
5. Nezlin M. V., Chernikov T. P., *Journal "Physica Plasmy"*, 21, p. 975 (1995)
6. Goncharov A. N., Dobrovolsky A. N., Zatugan A. V., Procenko I. M., IEEE Transaction on Plasma Science, 21, p. 40 (1993)
7. Goncharov A. A., Dobrovolsky A. N., Kocarenko A. N., Morozov A. I., Procenko I. M., *Journal "Physica Plasmy"*, 20, p. 499 (1994)
8. Zadoroshny V. F., *The problems of physics cinetic*, Kyiv: "Naukova Dumka", 1990
9. Larichev V. D., Reznik G. N., *Journal "DAN USSR"*, 231, p. 1077 (1976)
10. Kamenetc F. F., Lahin V. P., Michajlovskij A. B., *Journal "Physica Plasmy"*, 13, p. 412 (1987)

ESTIMATES FOR LONG–TERM STABILITY FOR THE LHC

M. Böge and F. Schmidt *
CERN
CH-1211 Geneva 23

Abstract

Since about 10 years survival plots have been used to evaluate single–particle long–term stability. In a recent paper (M. Giovannozzi et al.) this concept has been reviewed, using a dynamic aperture (Dyn.Aper.) definition based on the average over different ratios of emittances. It has been shown that the survival times evaluated according to this procedure decay with the inverse of the logarithm of the number of turns in several different systems. In this paper the validity of this conjecture is tested in the case of the latest LHC lattice which has been studied extensively.

The inverse log conjecture also predicts a non–zero Dyn.Aper. at infinite times called D_∞. The tracking data are analysed for the LHC lattice to determine the relation between D_∞ and the onset of chaos determined through Lyapunov exponents. Two different methods to automate the prediction of the Lyapunov exponent are tested and are compared with D_∞.

1 Introduction

In Ref. [1] (see also Ref. [2]) it has been shown that for several dynamical systems the evolution of the Dyn.Aper. $D(N)$ as a function of turn number N is well described by the following equation here called the *Inverse Log Conjecture*:

$$D(N) = D_\infty \left(1 + \frac{b}{log_{10}(N)}\right). \tag{1}$$

The D_∞ can be interpreted as the Dyn.Aper. after an infinite number of turns while the b appears to be a measure of the range of amplitudes where particle loss will take place, e.g. a value $b = 3$ means that after 1'000 turns the Dyn.Aper. is still a factor of two larger than D_∞. For this relation to work a precondition is to average the Dyn.Aper. over the four dimensional phase space as described in Ref. [3]:

$$D(N) = \left(\int_0^{\pi/2} [D_\alpha(N)]^4 \sin(2\alpha) d\alpha\right)^{1/4}, \tag{2}$$

where α is related to emittance ratio ϵ_{II}/ϵ_I by:

$$\alpha = atan\sqrt{\epsilon_{II}/\epsilon_I}, \tag{3}$$

---

*This research was supported in part by the National Science Foundation under Grant No. PHY94-07194.

e.g. ($\alpha = 45°$) corresponds to a emittance ratio of ($\epsilon_{II}/\epsilon_I = 1$). As the tracking for the LHC is usually done in the full six dimensional phase space one could argue that an average over the six dimensions is needed. This is not done for the following reasons: firstly the nonlinear coupling between longitudinal and transverse planes is small which allows the separate treatment of the longitudinal plane, secondly for the LHC tracking the initial conditions in the longitudinal phase space are not varied but fixed to one set of pessimistic (large) values and lastly the tracking effort would have to be increased by another factor of ten. One aim of this report is to check the conjecture for the LHC version 4 which has been extensively studied (see Ref.[4]). Another aim is the understanding of the relation between D_∞ and the onset of chaos.

2 Fitting Technique

One can rewrite Eq. 1 as follows:

$$D(N) \cdot \log_{10}(N) = D_\infty \cdot \log_{10}(N) + D_\infty \cdot b, \tag{4}$$

where $\log_{10}(N)$ is treated as an independent variable on the right–hand side. Thus D_∞ denotes the slope and $D_\infty \cdot b$ the offset of a linear function which describes $D(N) \cdot \log_{10}(N)$. A linear regression yields both quantities with a certain error Δ. The error of $D(N)$ is calculated to be:

$$\Delta(D(N)) = \Delta(D_\infty) + \Delta(D_\infty \cdot b)\frac{1}{\log_{10}(N)} \tag{5}$$

It should be noted that the multiplication of $D(N)$ with $\log_{10}(N)$ in Eq. 4 puts a stronger weight on loss boundaries where they are most relevant, i.e. at larger turn numbers N.

3 Conjecture Test

Figure 1 summarises the tracking data and the fitting result for one realization of the imperfect LHC: the tracking has been performed for 17 emittance ratios up to 10^6 turns. For the emittance ratio of one ($\alpha = 45°$) the tracking has been prolonged to 10^7 turns. A linear regression fit according to Eq. 4 is performed up to 10^5 and 10^6 turns. The fits are extrapolated to 10^7 turns and quoted with their errors. The data for $\alpha = 45°$ which deviate from the phase space averaged data at small turn numbers are consistent with both fits beyond 10^6 turns within their errors. Moreover, reducing the number of angles to 9 changes the predicted D_∞ by a mere 1.1%.

A bit worrying is the fact (see Figure 2) that both the fitted D_∞ and the error of the fit are increasing between 10^5 and 10^6 turns. The figure shows that this is due to a monotonically increasing sliding fit of D_∞ after a few thousand of turns, i.e. the Dyn.Aper. decreases less rapidly than the linear fit does imply. Using the conjecture fit for 60 machine realization (Figure 3) reveals a small anti correlation between D_∞ and b which could mean that the linear relation of Eq. 4 is based on a too simple assumption. On the other hand the figure also shows that the fit constants and the Dyn.Aper. scaled from 10^5 to 10^6, using the inverse log conjecture, have small errors. Even though the significance of the fit parameters remains unclear the fit with two parameters may still be useful to extrapolate the Dyn.Aper. to larger turn numbers.

To check this assumption emittance ratio scans have been extended up to 10^6 turns for 5 different realizations of the random errors (see Figure 4). The fit involving data up to 10^5 turns and the tracking data for 10^6 turns agree within the error bars of the extrapolation.

4 Chaos and D_∞

Since many years the chaotic boundary has been used to estimate the long-term Dyn.Aper. (see Ref. [5]). D_∞ determined from the conjecture fit should agree with the onset of chaos because both quantities describe the stability boundary in phase space. Agreement of the two independent methods would give D_∞ a physical meaning at least in a heuristic manner.

It is well known that there cannot be a rigorous non–zero loss boundary over infinite number of turns in a system with more than two degrees of freedom due to the loss of particles in the Arnold web (see Ref.[6]). However tracking studies for various systems have clearly shown that there always seems to be a hard core of stability in the amplitude space which is equivalent to a non–zero D_∞.

Two models have been tested: the four dimensional Hénon model and the LHC case for which the conjecture fit is shown in Figure 1. Due to its simplicity the first model can be tracked for a large number of angles and turn numbers (40 and 10^7 respectively). The LHC has been tracked for only 17 angles and 10^6 turns which has required two weeks of CPU time of a powerful 10 processor workstation cluster [7].

The two models are meant to be independent: the LHC is tracked at its nominal set of tunes (Q_x=63.28, Q_y=63.31) while the tunes of the Hénon model are chosen so as to maximise the chaotic regime (Q_x=0.168, Q_y=0.201) [8]. The *Top Left* and *Bottom Left* part of figure 5 depict the Dyn.Aper. versus emittance ratio of the Hénon and LHC model respectively, each curve corresponding to a different number of turns. The variations of the Dyn.Aper. can be large and depend on the choice of the tunes and on the realization of the random errors. For the phase space averaged Dyn.Aper. the conjecture fit agrees well with the tracking data in both cases (see *Top Right* and *Bottom Right* of Figure 5).

Chaos is detected by tracing the path of two initially close–by particles. This method is preferred over the original one introduced by Benettin et al. [9] as in this context the most sensitive measure is more relevant than the precise knowledge of the Lyapunov exponent.

Owing to the fact that the automatic detection of the onset of chaos is much more difficult than the reliable but time consuming inspection by eye a new approach has been attempted. Two different values can be automatically extracted from the tracking data: in the first method the distance in phase space must exceed a threshold which is larger than the final separation of any two regular (initially close–by) particles at the end of the tracking, whereas in the second method the motion is deemed chaotic when the slope, calculated from the evolution of the distance in phase space in a double logarithmic scale, is outside a certain interval of slope values (for regular motion the slope is one). In the following these techniques are called the **distance** and the **slope** method respectively. For simplicity the threshold and the slope interval are kept constant but it may be advantageous to vary them as a function of turn number.

The distance method is certainly safe due to its definition. However, it is an optimistic estimate because weakly chaotic particles may not have enough time to separate beyond the chosen threshold. The slope method is less precisely defined: it may be optimistic in case the motion is so weakly chaotic that the slope is not affected but it could also be pessimistic because it can pick up large oscillations of particles which are close to some resonance but which are nevertheless regular. The slope method is chosen as the preferred indicator as it is usually more pessimistic and more consistent with the inspection by eye. It should be mentioned that both methods can be improved by using frequency analysis [10] which allows to eliminate most of the regular oscillations from the evolution of the distance in phase space.

In the case of the Hénon model the slope method is pessimistic (Top right in Figure 5) and very close to the D_∞ fit. From the above discussion it is not surprising that D_∞ itself varies widely as a function of turns. As expected the distance method is optimistic at low turn numbers. At 10^7 turns, however, all three curves converge to almost the same point. It should be noted that this behaviour has been reproduced at two other tune working points. Tab. 1 summarises the results for all three cases.

Table 1: Results for the Hénon Model

Turn Number	Q_x	Q_y	D_∞	Chaos (Slope Method)	Chaos (Distance Method)
10^5	0.168	0.201	114.6	118.1	122.3
10^7			119.7	121.8	120.0
10^5	0.201	0.168	–	123.7	127.3
10^7			125.3	125.1	125.1
10^5	0.201	0.112	–	69.8	73.6
10^7			72.7	72.9	71.7

For the LHC the distance and the slope method are both optimistic (Bottom right in Figure 5). The fact that the slope method is optimistic means that the Dyn.Aper. of the LHC is determined by very weak chaotic motion. Still, at large turn numbers the chaotic boundary determined by the latter method agrees quite well with D_∞.

In both models the fit of D_∞ appears to be pessimistic in an intermediate turn number regime. In fact, in all studied LHC cases D_∞ is a too pessimistic estimate of long–term stability.

5 Conclusion

The inverse log conjecture has been thoroughly tested for the LHC version 4. Although doubts remain about the physical meaning of D_∞ and b the fit can be used to extrapolate the Dyn.Aper. from 10^5 to 10^6 turns. There are indications that this extrapolation can be further extended to 10^7 turns.

The chaos and D_∞ border seem to converge for large turn numbers for both the Hénon and the LHC model.

6 Acknowledgements

We would like to thank E. Todesco for discussing the conjecture fit for the Hénon model with one of the authors at the Santa Barbara workshop.

References

[1] M. Giovannozzi, W. Scandale and E. Todesco, "Prediction of long–term stability in large hadron colliders", CERN LHC Project Report **45/rev** (1996) and Part. Accel., in press;
M. Giovannozzi, W. Scandale and E. Todesco, "Inverse logarithmic extrapolation of survival plots in hadron colliders", Beam Dynamics Newsletter **12** (1996).

[2] M. Giovannozzi, W. Scandale and E. Todesco, "Inverse logarithm decaying of long–term dynamic aperture in hadron colliders", presented at the Particle Accelerator Conference, Vancouver, 12–16 May, (1997).

[3] E. Todesco and M. Giovannozzi, "Dynamic aperture estimates and phase space distortions in nonlinear betatronic motion", Phys. Rev. E **53**, (1996) 4067.

[4] M. Böge and F. Schmidt, "Tracking studies for the LHC optics version 4 at injection energy", presented at the Particle Accelerator Conference, Vancouver, 12–16 May, (1997), http://wwwslap.cern.ch/frs/report/dyn97pac.ps.Z.

[5] F. Schmidt, F. Willeke and F. Zimmermann, "Comparison of methods to determine long–term stability in proton storage rings", Part. Accel. **35**, pp. 249–256 (1991), http://wwwslap.cern.ch/frs/report/lyasl.ps.Z.

[6] B.V. Chirikov, "A universal instability of many–dimensional oscillator systems", Physics Reports **52**, pp. 263–379, (1979).

[7] E. McIntosh, T. Pettersson and F. Schmidt, "A Proposal for a Numerical Accelerator Project", presented at the "LHC95 International Workshop on Single–particle Effects in Large Hadron Colliders", Montreux, October 1995, http://wwwslap.cern.ch/frs/report/petter.ps.Z.

[8] E. Todesco, private communication.

[9] G. Benettin, L. Galgani and J.M. Strelcyn, "Kolmogorov Entropy and Numerical Experiments", Phys. Rev. A **14**, (1976) 2338.

[10] R. Bartolini and F. Schmidt, "Evaluation of Non-Linear Phase Space Distortions via Frequency Analysis", LHC Project Report 98 and in the proceedings of the workshop on: "Nonlinear and Collective Phenomena in Beam Physics", Arcidosso, September 1996, http://wwwslap.cern.ch/frs/report/arcif.ps.Z.

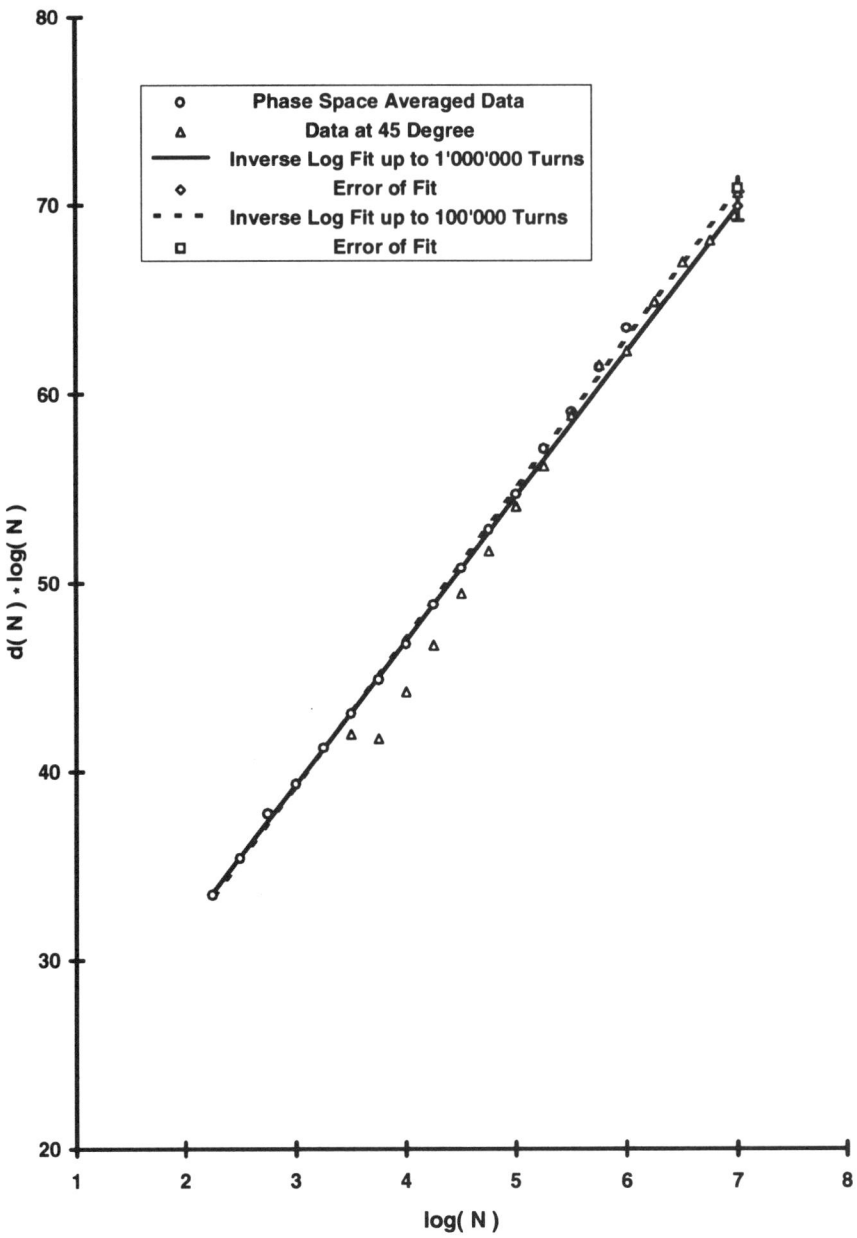

Figure 1: *Fits of Eq. 4 from 10^2 to 10^5 and 10^6 as well as the extrapolation to 10^7 turns for one realization of the imperfect LHC*

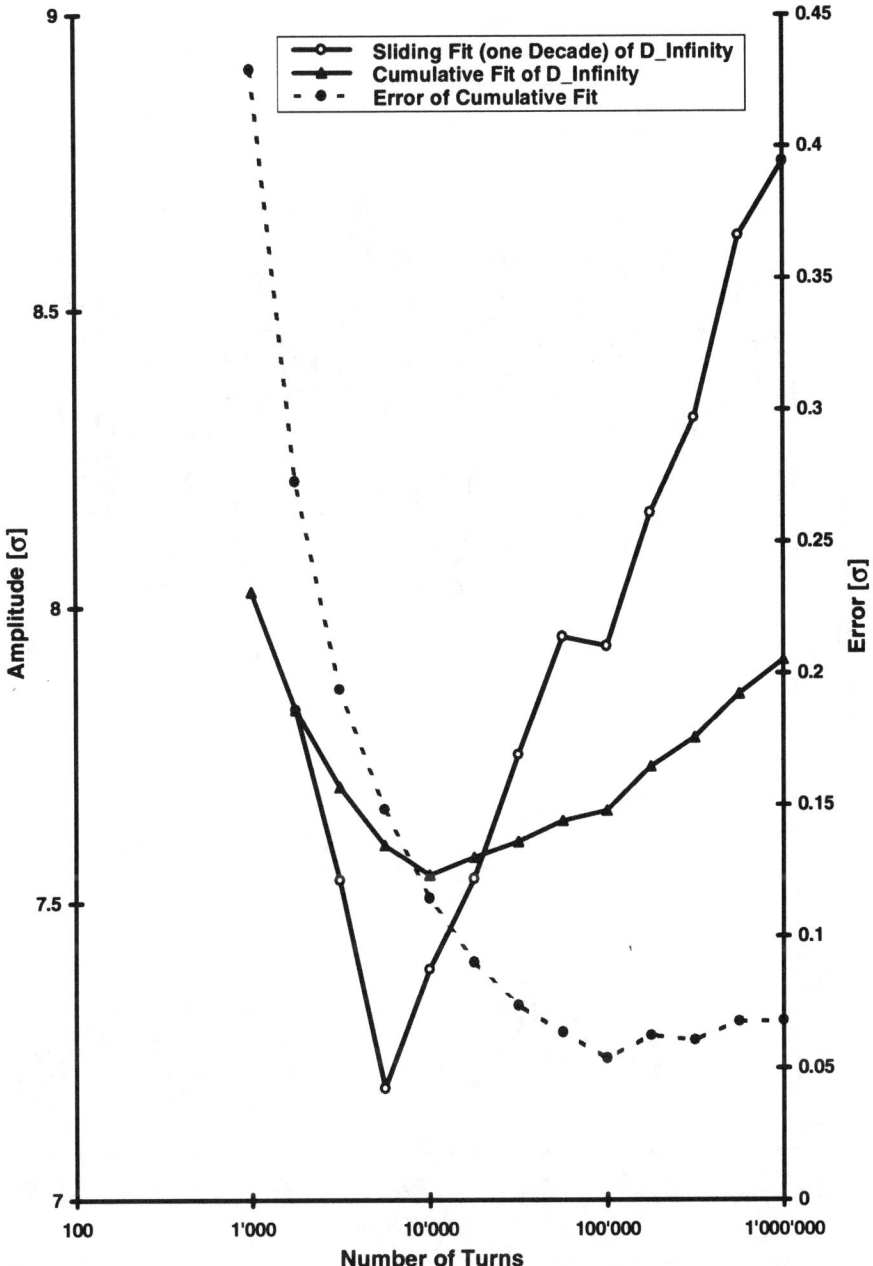

Figure 2: D_∞ determined from a cumulative fit and a sliding fit

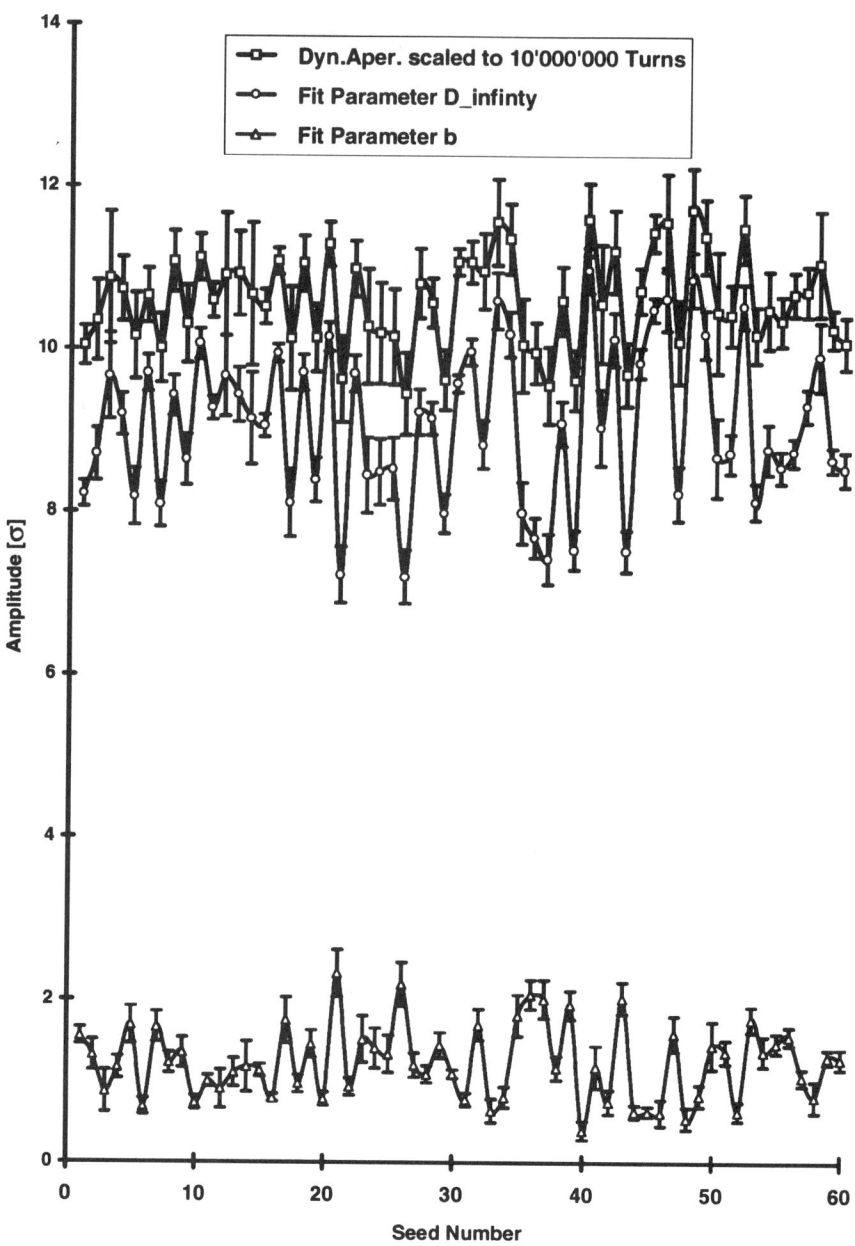

Figure 3: *Scaled Dyn.Aper. and the conjecture fit parameters D_∞ and b*

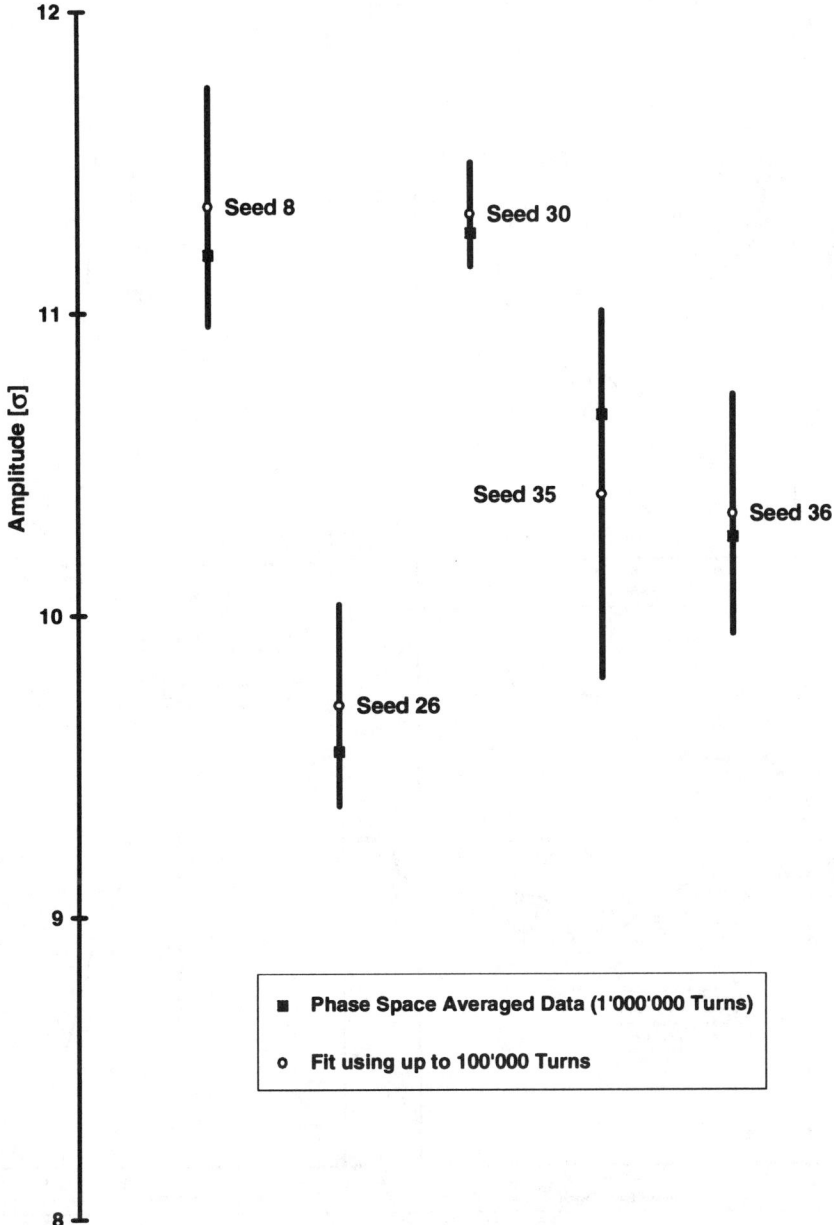

Figure 4: *Comparison of tracked and scaled Dyn.Aper.*

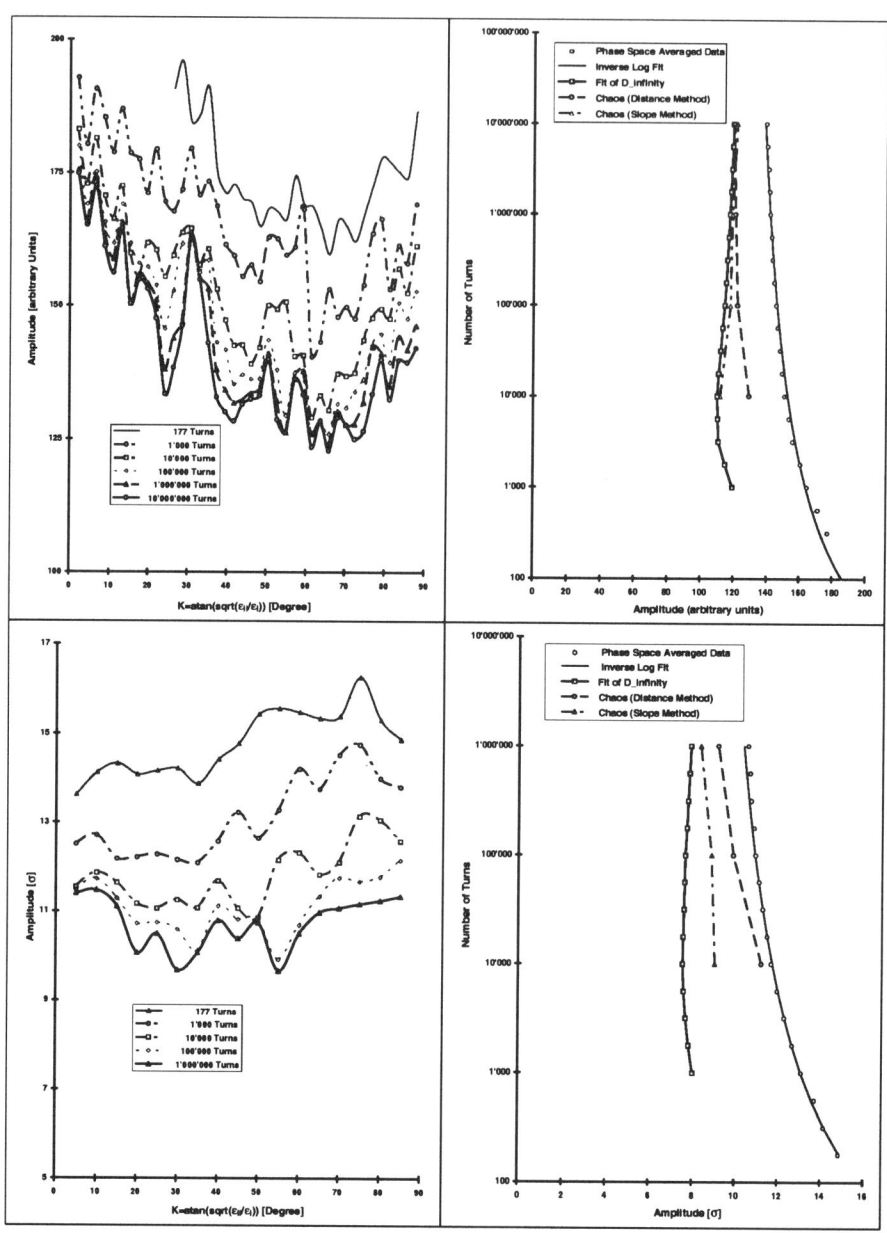

Figure 5: *The Hénon model – Top Left: Stable amplitude versus emittance ratio between 10^2 and 10^7 turns, Top Right: Survival plot, conjecture fit and chaos boundaries,*
The LHC model – Bottom Left: Stable amplitude versus emittance ratio between 10^2 and 10^6 turns, Bottom Right: Survival plot, conjecture fit and chaos boundaries

Halo Control, Beam Matching, and New Dynamical Variables for Beam Distributions

W. Lysenko* and Z. Parsa[†]

*Los Alamos National Laboratory, MS H808, Los Alamos, NM 87545, USA
[†]Brookhaven National Laboratory, 901A Physics Dept., Upton, NY 11973-5000, USA

Abstract.
We present the status of our work on physics models that relate to the understanding and control of beam halo, which is a cause of particle loss in high power ion linear accelerators. We can minimize these particle losses, even in the presence of nonlinearities, by ensuring the beam is matched to high order. Our goal is to determine new dynamical variables that enable us to more directly solve for the evolution of the halo. We considered moments and several new variables, using a Lie-Poisson formulation whenever possible. Using symbolic techniques, we computed high-order matches and mode invariants (analogs of moment invariants) in the new variables. A promising new development is that of the variables we call weighted moments, which allow us to compute high-order nonlinear effects (like halos) while making use of well-developed existing results and computational techniques developed for studying first order effects.

I INTRODUCTION

For a high power ion linear accelerator, we require that the beam be deposited on the target with minimal beam losses in the accelerator. In such applications, we are not interested in any details of the single-particle motion but in the evolution of the phase-space distribution. In this paper, we consider various ways to compute beam behavior in terms of distributions. Such approaches are particularly appropriate to high power ion linacs because we need to consider the distribution anyway to take into account the rather substantial space-charge forces.

In the distribution approach, the central idea is that of matching the beam to the accelerator structure. Concepts such as aberrations (nonlinearities in the map), which relate to the single-particle motion and are important in imaging applications, are not as relevant. When this single-particle motion is important, we often achieve the desired behavior by minimizing aberrations by introducing symmetries into the machine. In matching, however, the emphasis is more on the beam rather than the machine. The idea of matching is to introduce symmetries into the beam to get the desired behavior. In our case, the way to avoid halo formation and the associated particle loss in a linac is to keep the beam accurately matched (to high order) throughout the machine. In principle, we can eliminate emittance growth, halo formation, and particle loss even in a machine with aberrations by making sure the beam is exactly matched.

We will consider the evolution of the phase-space distribution in the Vlasov regime. To do this, we need dynamical variables to describe the distribution. Just as the coordinates x and p describe the dynamical state of a particle, moments of the phase-space distribution are an example of variables that describe the distribution. Moments are useful because they are closely related to laboratory quantities, such as rms beam sizes. We can compute the values of the moments of a beam at any time from their initial values using the moment-evolution equations. These evolution equations can be formulated as a Lie-Poisson system, which enforces Hamiltonian structure even though there are no Hamilton's equations for moment evolution. It is possible to compute matched values of the moments for given situations and to determine moment invariants, which are functions of moments (like rms emittance) that are conserved for linear motion.

The idea behind moments is that we solve directly for the quantities of interest (properties of the distribution like beam sizes) rather than try to get them from the single-particle motion. Our goal is to find other variables that are even more suitable for studying the halo problem. We would like to solve directly for the evolution of the halo, *i.e.*, we are seeking some kind of "halo variables." For simplicity, the examples in this paper are for just one degree of freedom and without space charge. However, the concepts discussed are valid for three degrees of freedom (six phase-space dimensions) and space-charge forces.

II BEAM MATCHING

A matched beam is one whose distribution function $f(x, p, t)$ is a function of single-particle invariants. This definition is consistent with the usual ones but is also useful when we have nonlinearities. For a linear periodic lattice, for example, $f(x, p, t)$ is matched if it is an arbitrary function F of the Courant-Snyder invariant ellipse

$$f(x, p, t) = F(\gamma x^2 + 2\alpha x p + \beta p^2). \tag{1}$$

This just says that, for a collection of ellipses all having the shape of the invariant ellipse, the phase-space density must be constant along any ellipse but may vary from ellipse to ellipse. Thus we may have both a hollow beam and also one peaked in the center that are both exactly matched to a given beamline. Any matched distribution function is periodic in time, as are all its moments. If we see that one of the moments does not have the same value at times one lattice period apart, we know the beam is not matched.

It is this property that makes the use of moments (or other distribution variables) so useful for computing matches and studying mismatch effects. The problem with the single-particle motion is that it always contains a betatron-frequency component, even in a matched beam. It is not possible to say anything about whether the beam is matched or not by looking at any single-particle trajectory. A matched beam has the symmetry required to ensure that it contains only the frequencies present in the Hamiltonian. In the present case of a linear period lattice, the Hamiltonian contains the frequency of the lattice and its harmonics. The betatron frequency, although present in the single-particle motion and in general (mismatched) distributions, is not present in matched distributions. It is averaged away because of the elliptical symmetry of the matched distributions.

To compute a beam matched to a periodic system, linear or not, we simply need to find a set of moments that are periodic. If the forces do not depend on time (this situation is often called smooth or continuous focusing, as distinguished from situations like alternating-gradient focusing), then matched distributions are constant in time. Matched moments in this case are time independent.

III HALO VARIABLES

An ideal set of variables would be one that includes a variable that describes the fraction of the beam outside a given radius. Knowing the evolution equation for such a variable would allow us to compute and to be able to control particle loss. How close can we come to this goal?

Here, we will begin to answer this question by considering several different descriptions of beam distributions. We will compute matched beams for a simple nonlinear example and, where possible, compute mode invariants (the analog of moment invariants) in the new variables. One question we would like to answer is why no one has been able to determine the moment invariants for nonlinear motion. Is this just a problem for moments or is it a more general situation?

IV LIE-POISSON FORMULATION

We can maintain Hamiltonian structure in distribution coordinates as follows[1]. Suppose we describe the solution to the Vlasov equation with some

---

[1] We thank D. Holm for pointing out this approach (private communication, 1989).

dynamical variables G_i. Think of the G_i as functionals that map the phase-space distribution function $f(x,p,t)$ to numbers (the moments, e.g., in a moments description). The variables G_i will form a Lie algebra if we define a Lie-Poisson (LP) bracket in terms of the ordinary Poisson bracket by

$$[G_i, G_j]_{\text{LP}} = \iint dx\, dp\, f(x,p,t) [\frac{\delta G_i}{\delta f}, \frac{\delta G_j}{\delta f}], \tag{2}$$

where $\delta G/\delta f$ is the variational derivative of G with respect to f. The dynamics (evolution of the variables) is given by

$$\frac{d}{dt} G = [G, H_{\text{LP}}]_{\text{LP}}. \tag{3}$$

The LP Hamiltonian H_{LP} is derived from the usual single-particle Hamiltonian H by

$$\frac{\delta H_{\text{LP}}}{\delta f} = H. \tag{4}$$

If the functional G is an integral over phase space involving f, x, p, and t, then the variational derivative is the partial derivative of the integrand with respect to f. In this case,

$$H_{\text{LP}} = \iint dx\, dp\, f(x,p,t) H(x,p,t). \tag{5}$$

V EXAMPLES OF NEW VARIABLES IN LIE-POISSON FORMULATION

We compute matched beams and mode invariants symbolically using *Mathematica* [1] for three kinds of variables: moments, Fourier modes, and what we call histogram modes. For each of these variables, we first define the properties of the LP bracket and determine the LP Hamiltonian, H_{LP}. We can then compute the time derivative of any quantity by taking the bracket of the quantity with H_{LP}. In these examples, we consider the following single-particle Hamiltonian

$$H(x,p) = \frac{p^2}{2} + k\frac{x^2}{2} + k_2 \frac{x^3}{3}. \tag{6}$$

We assume the force constants k and k_2 are time-independent. In this case, matched beams are those whose distributions are constant in time. In these examples, we are also neglecting space charge. To include space charge, the force constants would have to depend on the spatial moments (moments involving x).

We compute matches by requiring the LP bracket of all distribution variables (all moments, for example) with H_{LP} to be zero. To compute mode invariants, we assume some function of the variables and solve for parameters of the function that result in making the function time independent.

Moments

Moments are averages of monomials in phase space over the phase-space distribution. For example, $<x^2>$ is defined for continuous and discrete distributions by

$$<x^2> = \iint dx\, dp\, f(x,p,t)\, x^2 = \frac{1}{N} \sum_{i=1}^{N} x_i^2. \tag{7}$$

We get the LP bracket for moments from Eq. 2

$$[<x^i p^j>, <x^m p^n>]_{\text{LP}} = (in - jm)<x^{i+m-1} p^{j+n-1}>. \tag{8}$$

We achieve closure for finite order by replacing the above right hand side with zero if it involves moments of orders greater than the cutoff value. Note that, given the moments, it is difficult to compute the distribution function. Fortunately, we do not have to do this to compute the LP Hamiltonian. Actually, from Eq. 5, we see that, since the single-particle Hamiltonian (Eq. 6) is a polynomial, there is nothing to compute. We just wrap brackets around all the terms in the single-particle Hamiltonian to get the LP Hamiltonian.

$$H_{LP} = \frac{<p^2>}{2} + k\frac{<x^2>}{2} + k_2\frac{<x^3>}{3}. \tag{9}$$

Matched moments

Let us truncate at fourth order, zeroing all moments of order five and above. We compute the matched set of moments by requiring that the LP bracket of all moments with the LP Hamiltonian be zero. Applying this condition results in the following relations.

$$\begin{aligned}
&<p^2> = k<x^2> - \frac{5k_2^2}{k^2}<x^2 p^2> &&<p^3> = 0 \\
&<xp> = 0 &&<x^4> = \frac{3<x^2 p^2>}{k} \\
&<x^3> = -\frac{5k_2}{k^2}<x^2 p^2> &&<x^3 p> = 0 \\
&<x^2 p> = 0 &&<xp^3> = 0 \\
&<xp^2> = -\frac{k_2}{k}<x^2 p^2> &&<p^4> = 3k<x^2 p^2>
\end{aligned} \tag{10}$$

Any set of moments related this way will be constant in time. There are two things to notice here. First, the nonlinearity (term involving k_2 in the Hamiltonian) give rises to a nonzero value of some moments that are of odd degree in x. A distribution matched to a nonlinear system does not have elliptical symmetry. Second, we see that, even in the absence of the nonlinearity (*i.e.*, for $k_2 = 0$), the higher moments (here fourth) must have certain values for the beam to be exactly matched. It is not enough to set just the second moments correctly to get an exactly matched beam.

Since there are twelve moments of orders two through four and there are ten matching conditions (Eq. 10), a beam matched to fourth order has two free parameters. In the case of linear motion, it is useful to take as these parameters the rms emittance and the fourth-order emittance, which we define below.

Moment invariants

Moment invariants are functions of moments that are conserved in time. For a fourth-order cutoff, we determine these invariants by assuming they are fourth-degree polynomials in the moments of orders two through four

$$I = c_1 <x^2> + c_2 <xp> + \cdots + c_{1819} <p^4>^4. \tag{11}$$

There are 1819 terms in this expression. Because there are so many terms, we must use a computer algebra system like *Mathematica* to do such calculations. We need to determine the values of c_i that make I constant in time. To do this, we take the LP bracket of I with the LP Hamiltonian and set the result to zero. This is actually a number of conditions, since the invariant should be constant for any beam, *i.e.*, any set of moments, and for any values of the force constants. Taking this into account, we determine most of the c_i values in terms of the remaining ones (the number remaining is equal to the number of functionally independent invariants). Substituting these values into Eq. 11 yields the invariant I, which is actually a sum of several independent invariants and functions of invariants. The *Mathematica* code finds the complete set of functionally-independent invariants, including those of mixed order (*e.g.*, containing both second and fourth moments).

For the linear case ($k_2 = 0$), we find the invariant

$$\epsilon_2^2 = <x^2><p^2> - <xp>^2, \tag{12}$$

which is just the square of the usual rms emittance, which we write as ϵ_2 because it is a function of second moments. We also get higher-order invariants like

$$\epsilon_4^4 = <x^4><p^4> - 4<x^3p><xp^3> + 3<x^2p^2>^2, \tag{13}$$

which we already know about (see the review in Ref. [2]). We can consider ϵ_4 to be a fourth-order emittance since it has the emittance-like property that the expression given by Eq 13 is always nonnegative, for any set of moments corresponding to a beam with a nonnegative distribution function. Moment invariants for nonlinear motion are still unknown. (Invariants involving only moments of the cutoff order, like ϵ_4, are a special case; they are invariant even for nonlinear motion in the truncated system.)

The matched values of the second moments can be expressed in terms of the Courant-Snyder parameters and the rms emittance by

$$<x^2> = \beta\epsilon_2$$
$$<xp> = -\alpha\epsilon_2 \qquad (14)$$
$$<p^2> = \gamma\epsilon_2$$

The relations of Eq. 14 have analogs for higher orders. We can relate the matched values of the fourth moments to the Courant-Snyder parameters and the fourth order emittance. For example, we have $<x^4> = (\sqrt{3}/2)\beta^2\epsilon_4^2$ and there are similar expressions for the other four fourth moments.

Fourier Modes

Define Fourier modes on a finite region of phase space by

$$f_{mn} = \frac{1}{(2\pi)^2} \int_{-\pi}^{\pi} dx \int_{-\pi}^{\pi} dp\, f(x,p,t)\, e^{-i(mx+np)} \qquad (15)$$

The LP bracket of two of these quantities is

$$[f_{ij}, f_{mn}]_{\text{LP}} = -\frac{1}{(2\pi)^2}(in - jm)\, f_{i+m,j+n}. \qquad (16)$$

We attempt closure by zeroing higher modes. Unlike the situation for moments, this does not preserve all the properties of a Lie algebra for the truncated system, but we proceed anyway, in the hope it will lead to something useful. (The problem is that the LP bracket of two modes with orders higher than the cutoff may yield a low-order mode. In some cases, this means the Jacobi identity is not satisfied.) The LP Hamiltonian can be computed by expressing the distribution function in terms of the modes by

$$f(x,p,t) = \sum_{m,n=-N}^{N} f_{mn}\, e^{i(mx+np)} \qquad (17)$$

and then using Eqs. 5 and 6, doing the integral symbolically.

Matched Fourier modes

For a cutoff of $N = 1$, the matched modes satisfy

$$f_{-1,0} = (k+\bar{k})f_{0,-1} \qquad f_{-1,-1} = \frac{k+\bar{k}}{k-\bar{k}}f_{1,1}$$
$$f_{1,0} = (k-\bar{k})f_{0,-1} \qquad f_{-1,1} = \frac{k+\bar{k}}{k-\bar{k}}f_{1,1} \qquad (18)$$
$$f_{0,1} = f_{0,-1} \qquad f_{1,-1} = f_{1,1}$$

where $\bar{k} = ik_2(\pi^2 - 6)/3$. There are nine modes at this order, so there are three free modes for a matched beam.

Fourier mode invariants

For linear motion ($k_2 = 0$), some invariants (truncated at order 1) are

$$\begin{aligned}
&f_{0,0} = \text{const.} \\
&f_{-1,-1} + f_{-1,1} + f_{1,-1} + f_{1,1} = \text{const.} \\
&f_{0,-1}f_{0,1} + f_{-1,1}f_{1,-1} + f_{-1,0}f_{1,0} + f_{-1,-1}f_{1,1} = \text{const.}
\end{aligned} \qquad (19)$$

The first of these is just the conservation of particles and is equivalent to $<1>=$const. for moments. With the nonlinearity turned on, we see that the second of Eq. 19 is no longer an invariant but the other two expressions are still invariant for the truncated system.

Histogram Variables

Let us divide phase space into rectangular bins labeled by the indices m in the x-direction and n in the p-direction. Let both indices range from $-N$ to N ($4N^2$ bins total). Let the mode f_{mn} describe the density in phase space at the bin specified by the indices m and n. The following LP bracket will approximate the correct physics (again, the truncated system is not exactly Hamiltonian).

$$[f_{ij}, f_{mn}]_{\text{LP}} = \begin{cases} (m-i)(n-j)(f_{in} - f_{mj}), \\ \qquad |m-i| = 1 \text{ and } |n-j| = 1 \\ 0, \qquad \text{otherwise.} \end{cases} \qquad (20)$$

Note that closure for these variables is a boundary problem. If the beam is localized (as are real beams) then closure is automatic and involves no additional physics approximations. Increasing the order of the approximation in this scheme amounts to adding more bins to the periphery of phase space, which does not affect any results if the beam itself is truncated (does not extend past the boundary). The LP Hamiltonian is

$$H_{\text{LP}} = \sum_{m,n=-N}^{N} f_{mn} <H>_{mn}, \qquad (21)$$

where $<H>_{mn}$ is the average of $H(x,p)$ over the bin.

Matched histogram modes

To compute a matched beam for histogram modes, we require every histogram value to be constant in time. We find N independent modes (bin values) in a matched beam, out of a total of $4N^2$ modes. Figure 1 shows an example with and without the nonlinearity. Equal densities in phase space are depicted by equal gray levels. In the continuous case, we know a matched

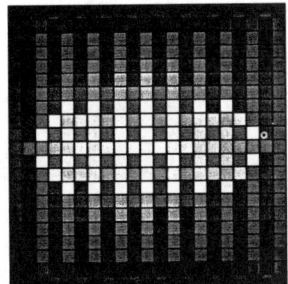 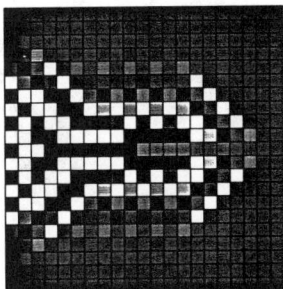

FIGURE 1. Matched histogram without nonlinearity (left) and with nonlinearity (right).

is elliptical for linear focusing forces. For the nonlinear case, we have the fish shape familiar from longitudinal motion in rf machines. The histogram matched modes preserve this basic structure.

Histogram mode invariants

For the nonlinear case, we found one invariant of degree 1 in the modes. For N odd it consists of the sum of all modes having odd indices and for N even, it consists of the sum of all modes having even indices. This is approximately a statement that the total number of particles is conserved. For the linear motion case ($k_2 = 0$), we find, in addition, that the mode $f_{0,0}$ is separately conserved. We have not computed invariants of higher degree in the modes.

VI WEIGHTED MOMENTS

If we want to describe a beam accurately using moments, we need to include moments beyond second order. Higher-order emittances like ϵ_4, together with the usual rms emittance, can help us describe the halo of a beam. However, we would like a description that better factors the core and halo components. So let us consider generalizing the usual moments. Let us keep monomials in the phase-space variable x and p as the basis functions, but change the weights. Define the weighted moment of weight index n of the phase-space distribution $f(x, p, t)$ for continuous and discrete distributions by

TABLE 1. Courant-Snyder parameters determined from a beam using second moments of different weights.

	α	β	ϵ_2
core	0.200	1.000	0.200
n= 3	0.201	1.000	0.200
n= 2	0.197	1.005	0.202
n= 1	0.164	1.055	0.217
n= 0	0.011	1.360	0.316
n=−1	−0.153	1.839	0.515
n=−2	−0.193	1.976	0.589
n=−3	−0.198	1.992	0.599
halo	−0.200	2.000	0.600

$$<g(x,p)>_n = \frac{\iint dx\, dp\, [f(x,p,t)]^{n+1} g(x,p)}{\iint dx\, dp\, [f(x,p,t)]^{n+1}}$$

$$= \frac{\sum_{i=1}^{N} [f(x_i,p_i,t)]^n g(x_i,p_i)}{\sum_{i=1}^{N} [f(x_i,p_i,t)]^n}, \qquad (22)$$

where the basis functions $g(x,p)$ are monomials in x and p. The usual moments result when the weight index is $n=0$. Normalized distribution functions do not make sense here since the normalization constant depends on n.

For negative values of n, the halo is emphasized. Table 1 shows the Courant-Snyder parameters computed (using Eq. 14) from moments of different weights for an example beam consisting of 10^6 particles. The beam consists of two components, which we label as "core" and "halo," each consisting of uniformly-filled ellipse. The halo, whose phase-space density is one-tenth that of the core, contains 23% of the particles. The first and last rows of the table show the Courant-Snyder parameters of the core and halo components of the beam. We see that we need not go to very high weight indices to measure halo properties using weighted moments.

A LP formulation of weighted moments is difficult (part of the problem is that they are defined by a nonlinear functional) and will not be attempted here. A nice feature of the weighted moments is that we already know the dynamics of these objects. Since the distribution function is constant on phase-space trajectories, so is any function of it. Thus

$$\frac{d}{dt}<g(x,p)>_n = <\frac{d}{dt}g(x,p)>_n, \qquad (23)$$

just as for regular moments. So, we already know the linear moment invariants.

For example, we have

$$<x^2>_n <p^2>_n - <xp>_n^2 = \text{const.}, \qquad (24)$$

which is the weighted-moment analog of rms emittance.

Instead of using higher-order moments to achieve a more accurate beam description, we can combine second-order moments of various weights. In the absence of space charge, moments of different weights do not interact, *i.e.*, the evolution equation for moments of a given weight index depend only on moments of the same weight index. This means the core and halo evolve independently. Space charge introduces coupling through the force constants, which depend on the spatial moments of all weights. If we numerically integrate the weighted-moment evolution equations to simulate beams, we need a space-charge model to determine the force constants from the moments. One way to do this would be to consider a collection of uniformly-charged ellipsoids, one for each weight index n. We can compute an effective linear force constant f_n^{eff}, which determines the evolution of the moments of weight index n, from the moments using

$$<xF_x>_n = k_n^{\text{eff}} <x^2>_n, \qquad (25)$$

where F_x is the total force from all ellipsoids. This is an obvious generalization of the result of Sacherer, which is used in linear (envelope) codes like TRACE 3-D [3]. We can compute the ellipsoid (and resulting space-charge field) for each weight index using methods identical to those used in the existing code.

VII DISCUSSION

Moments have been used to describe beam distributions. Matched values of moments can be easily computed, even to high order for nonlinear systems. This is an important advantage of the moment approach. To investigate the possibility of variables even better suited for halo studies, we looked at Fourier modes and histogram variables. Matched beams for nonlinear systems can be calculated for these variables. We did not find nontrivial mode invariants (the analog of moment invariants) for nonlinear motion for these variables. Further study of these and other variables may be worthwhile. Improving the efficiency of the symbolic code should help in this work. We do not have high cutoff orders for Fourier modes nor mode invariants of degree higher than the first for histogram variables because of computational limitations.

Histogram variables are probably not the ultimate approach for halo studies but could be usefully further studied because they so completely separate the halo and core of the beam.

Weighted moments appear most promising for studying halo behavior. We can access high order features of the beam using just second moments. There is no need to go to higher moments to see the halo. Furthermore, this description is efficient. Second moments of only a few different weights need to be considered to effectively characterize both the core and halo of the beam. This has large practical benefits. We can use existing computational and analytic results, originally developed for studying first order effects, to study nonlinear beam dynamics. We believe that existing linear codes like TRACE 3-D can be easily modified to handle high order effects by considering second moments of different weights (*i.e.*, by describing the beam by a collection of sigma matrices of different weights).

ACKNOWLEDGEMENTS

We thank Paul Channell for helpful comments. This work was partially carried out at the Institute for Theoretical Physics, University of California, Santa Barbara, as part of the New Ideas for Particle Accelerators Program with partial support by the U.S. Department of Energy and the National Science Foundation.

REFERENCES

1. Wolfram, Stephen *The Mathematica Book*, 3rd ed., Champaign: Wolfram Media/Cambridge University Press, 1996.
2. W.P. Lysenko, "The moment approach to high-order accelerator beam optics," Nucl. Instr. and Meth. in Phys. Res. A 363 (1995) 90-99.
3. D.P. Rusthoi, W.P. Lysenko, and K.R. Crandall, "Further improvements in TRACE 3-D," (to be published in proceedings of the 1997 Particle Accelerator Conference).
4. W. Lysenko and Z. Parsa, "Beam Matching and Halo Control," (to be published in proceedings of the 1997 Particle Accelerator Conference).

Beam Stability and Nonlinear Dynamics
December 3–5, 1996

Coordinator: Z. Parsa

SCHEDULE

Tuesday, December 3, 1996:

Time:	Speaker:	Title:

Convener: Z. Parsa

8:00 am	*Registration*	*ITP Front Lobby*
8:40	J. Hartle, ITP Director Z. Parsa, BNL	Welcome Introduction & Welcome
9:00	J. Meiss, U Colorado	Single-Particle Hamiltonian Dynamics
9:45	*Refreshment Break*	*ITP Center Patio*

Convener: G. Guignard

10:15	J. Marsden, Cal Tech	Symplectic Geometry, Maps, Integrators
11:00	M. Berz, Michigan State	From Taylor Series to Taylor Models & Remainder Differential Algebra with Interval Arithmetic
11:45	A. Dragt, UM	Factorization of Taylor Maps
12:25 pm	*Lunch Break*	*ITP Center Patio*

Convener: J. Hagel

2:00	J. Irwin, SLAC	One-Turn Map Generators Aberration Sources, Consequences and Corrective Actions
2:40	E. Todesco, INFN	Long-Term Orbit Stability Predictions (Lypunov estimates, Normal Forms)
3:25	*Refreshment Break*	*ITP Center Patio*

Convener: W. Lysenko

4:00	J. Laskar, BDL	Frequency Map Analysis - Theory & Experiments
4:45	R. Warnock, SLAC	The Effect of Resonances on Long-Term Stability & Symplectic Full Turn Maps
5:30	*Wine & Cheese*	*ITP Center Patio*
6:15	*Conference Dinner*	*ITP Center Patio*

Wednesday, December 4, 1996:

Convener: D. Robin

8:30 am	V. Balandin, DESY	Nonlinear Spin Dynamics
9:10	F. Schmidt, CERN	Dynamic Aperture - Simulation and Experiment
9:55	*Refreshment Break*	*ITP Center Patio*

Convener: J. Ellison

10:25	F. Ruggiero, CERN	Longitudinal Beam Echoes & Diffusion Rates

11:05	R. Siemann, SLAC	Sawtooth Instability & Over-Shoot Phenomena
11:45	H. Yoshida, NAO	Instability of Periodic Orbit and Non-Integrability of Hamiltonian System
12:25 pm	*Lunch Break*	*ITP Center Patio*

Convener: J. Krommes

1:40	P.Zenkevitch, ITEP	Neutralized Beams: Landau Damping in Systems with Strong Nonlinearity
2:20	G. Guignard, CERN	Stability of Beams in a High Frequency Linac with Strong Wakefields
3:05	*Refreshment Break*	*ITP Center Patio*

Convener: E. Lessner

3:35	M. Zeitlin, IPME	Wavelet Analysis of Hamiltonian System and its Perturbations
4:20	S. Andriznov, St. Petersburg	Nonlinear Aberration Correction
4:55	S. Heifets, SLAC	Search of the Mechanism of the Saw-Tooth Instability
5:30	*Reception*	*ITP Center Patio*

Thursday, December 5, 1996:

Convener: A. Chao

8:30	F. Ruggiero, CERN	Collective Effects in LHC
9:05	J. Hagel, U. Maderia	Galaxy Dynamics: Resonance Analysis in Celestial Mechanics

9:50	*Refreshment Break*	*ITP Center Patio*

Convener: M. Berz

10:20	A. Pankin, INR	Nonlinear Structure Near Boundary of Marginal Stability
10:40	G. Stupakov	Nonlinear Dynamics of Single Bunch Instability in Accelerators
11:10	V. Zadorozhny, IC	The Dynamic 2-D Electron Beams of the Plasma Lense
11:30	TBA	Map Measurements
11:45	G. Guinard	Statistical Analysis of Emittance Growth
12:20 pm	*Lunch Break*	*ITP Center Patio*

Convener: R. Siemann

1:50	Round Table Discussions on Outstanding Issues in Beam Instabilities and Nonlinear Accelerator Dynamics	Z. Parsa, A. Dragt, G. Guignard J. Laskar, J. Meiss, E. Todesco, & other speakers and participants
3:00	Z. Parsa, BNL	Summary and Closing Talk

Conference Ends

LIST OF PARTICIPANTS*

Dan Abell	University of Maryland
Serge Andrianov	St. Petersburg State University
Vladimir Balandin	Michigan State University
Martin Berz	Michigan State University
Nathan Brown	G. H. Gillespie Associates, Inc.
Alexander Chao	Stanford Linear Accelerator Center
David Cline	University of California, Los Angeles
Alex Dragt	University of Maryland
James Ellison	University of New Mexico
Anna Fishchuk	Institute for Nuclear Research, Kiev
Vladimir Gorev	Kurchatov Institute
Gilbert Guignard	CERN, Switzerland
Johannes Hagel	Universidade da Madeira
Katherine Harkay	Argonne National Laboratory
Samuel Heifets	Stanford Linear Accelerator Center
John Irwin	Stanford Linear Accelerator Center
John Krommes	Princeton University
Jacques Laskar	Astronomie et Systemes Dynamiques
Guy Laval	Ecole Polytechnique
Eliane Lessner	Argonne National Laboratory
Walter Lysenko	Los Alamos National Laboratory
William Marciano	Brookhaven National Laboratory
Jerrold Marsden	California Institute of Technology
James Meiss	University of Colorado
Alexei Pankin	Institute for Nuclear Research, Kiev
Zohreh Parsa	Brookhaven National Laboratory
Alain Piquemal	CEA, Bruyers-le-Chatel
David Robin	Lawrence Berkeley National Laboratory
Francesco Ruggiero	CERN, Switzerland
Raymond Sawyer	University of California, Santa Barbara
Frank Schmidt	CERN, Switzerland
Robert Siemann	Stanford Linear Accelerator Center
Gennady Stupakov	Stanford Linear Accelerator Center
Andrei Terebilo	Stanford Linear Accelerator Center
Ezio Todesco	INFN Sezione di Bologna
Marco Venturini	University of Maryland
Chunxi Wang	Stanford Linear Accelerator Center
Robert Warnock	Stanford Linear Accelerator Center
Yiton Yan	Stanford Linear Accelerator Center
Haruo Yoshida	National Astronomical Observatory of Japan
V. Zadorozhny	National Academy of Sciences of Ukraine
Michael Zeitlin	Russian Academy of Sciences, St. Petersburg
Pavel Zenkevitch	Institute of Theoretical & Experimental Physics, Moscow

*This may not include all names, especially late registrants.

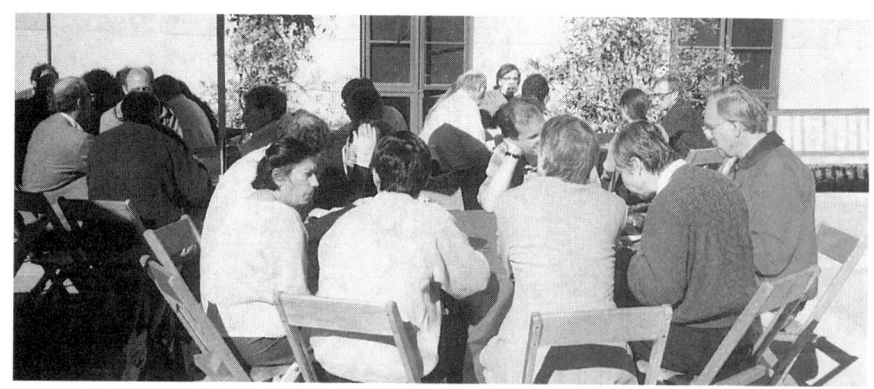

AUTHOR INDEX

A

Adrianov, S. N., 103

B

Balandin, V., 55
Berz, M., 1, 55
Böge, M., 201

D

Davydova, T. A., 149

E

Ellison, J. A., 41

F

Fedorova, A. N., 87

G

Giovannozzi, M., 25
Golubeva, N., 55
Goncharov, A. A., 189

H

Hagel, J., 133
Heifets, S., 117

I

Irwin, J., 173

L

Litovko, I. V., 189
Lysenko, W., 211

M

Mustafin, E., 75

O

Onishchenko, N., 189

P

Pankin, A. Yu., 149
Parsa, Z., 165, 183, 211

S

Scandale, W., 25
Schmidt, F., 63, 201

T

Todesco, E., 25, 63

W

Wang, C.-xi, 173
Warnock, R. L., 41

Z

Zadorozhny, V. F., 189
Zeitlin, M. G., 87
Zenkevich, P., 75, 165, 183

AIP Conference Proceedings

	Title	L.C. Number	ISBN
No. 247	Global Warming: Physics and Facts (Washington, DC 1991)	91-78423	0-88318-932-1
No. 248	Computer-Aided Statistical Physics (Taipei, Taiwan 1991)	91-78378	0-88318-942-9
No. 249	The Physics of Particle Accelerators (Upton, NY 1989, 1990)	92-52843	0-88318-789-2
No. 250	Towards a Unified Picture of Nuclear Dynamics (Nikko, Japan 1991)	92-70143	0-88318-951-8
No. 251	Superconductivity and its Applications (Buffalo, NY 1991)	92-52726	1-56396-016-8
No. 252	Accelerator Instrumentation (Newport News, VA 1991)	92-70356	0-88318-934-8
No. 253	High-Brightness Beams for Advanced Accelerator Applications (College Park, MD 1991)	92-52705	0-88318-947-X
No. 254	Testing the AGN Paradigm (College Park, MD 1991)	92-52780	1-56396-009-5
No. 255	Advanced Beam Dynamics Workshop on Effects of Errors in Accelerators, Their Diagnosis and Corrections (Corpus Christi, TX 1991)	92-52842	1-56396-006-0
No. 256	Slow Dynamics in Condensed Matter (Fukuoka, Japan 1991)	92-53120	0-88318-938-0
No. 257	Atomic Processes in Plasmas (Portland, ME 1991)	91-08105	0-88318-939-9
No. 258	Synchrotron Radiation and Dynamic Phenomena (Grenoble, France 1991)	92-53790	1-56396-008-7
No. 259	Future Directions in Nuclear Physics with 4π Gamma Detection Systems of the New Generation (Strasbourg, France 1991)	92-53222	0-88318-952-6
No. 260	Computational Quantum Physics (Nashville, TN 1991)	92-71777	0-88318-933-X
No. 261	Rare and Exclusive B&K Decays and Novel Flavor Factories (Santa Monica, CA 1991)	92-71873	1-56396-055-9
No. 262	Molecular Electronics—Science and Technology (St. Thomas, Virgin Islands 1991)	92-72210	1-56396-041-9
No. 263	Stress-Induced Phenomena in Metallization: First International Workshop (Ithaca, NY 1991)	92-72292	1-56396-082-6
No. 264	Particle Acceleration in Cosmic Plasmas (Newark, DE 1991)	92-73316	0-88318-948-8

Title	L.C. Number	ISBN
No. 265 Gamma-Ray Bursts (Huntsville, AL 1991)	92-73456	1-56396-018-4
No. 266 Group Theory in Physics (Cocoyoc, Morelos, Mexico 1991)	92-73457	1-56396-101-6
No. 267 Electromechanical Coupling of the Solar Atmosphere (Capri, Italy 1991)	92-82717	1-56396-110-5
No. 268 Photovoltaic Advanced Research & Development Project (Denver, CO 1992)	92-74159	1-56396-056-7
No. 269 CEBAF 1992 Summer Workshop (Newport News, VA 1992)	92-75403	1-56396-067-2
No. 270 Time Reversal—The Arthur Rich Memorial Symposium (Ann Arbor, MI 1991)	92-83852	1-56396-105-9
No. 271 Tenth Symposium Space Nuclear Power and Propulsion (Vols. I–III) (Albuquerque, NM 1993)	92-75162	1-56396-137-7 (set)
No. 272 Proceedings of the XXVI International Conference on High Energy Physics (Vols. I and II) (Dallas, TX 1992)	93-70412	1-56396-127-X (set)
No. 273 Superconductivity and Its Applications (Buffalo, NY 1992)	93-70502	1-56396-189-X
No. 274 VIth International Conference on the Physics of Highly Charged Ions (Manhattan, KS 1992)	93-70577	1-56396-102-4
No. 275 Atomic Physics 13 (Munich, Germany 1992)	93-70826	1-56396-057-5
No. 276 Very High Energy Cosmic-Ray Interactions: VIIth International Symposium (Ann Arbor, MI 1992)	93-71342	1-56396-038-9
No. 277 The World at Risk: Natural Hazards and Climate Change (Cambridge, MA 1992)	93-71333	1-56396-066-4
No. 278 Back to the Galaxy (College Park, MD 1992)	93-71543	1-56396-227-6
No. 279 Advanced Accelerator Concepts (Port Jefferson, NY 1992)	93-71773	1-56396-191-1
No. 280 Compton Gamma-Ray Observatory (St. Louis, MO 1992)	93-71830	1-56396-104-0
No. 281 Accelerator Instrumentation Fourth Annual Workshop (Berkeley, CA 1992)	93-072110	1-56396-190-3

Title	L.C. Number	ISBN
No. 282 Quantum 1/f Noise & Other Low Frequency Fluctuations in Electronic Devices (St. Louis, MO 1992)	93-072366	1-56396-252-7
No. 283 Earth and Space Science Information Systems (Pasadena, CA 1992)	93-072360	1-56396-094-X
No. 284 US-Japan Workshop on Ion Temperature Gradient-Driven Turbulent Transport (Austin, TX 1993)	93-72460	1-56396-221-7
No. 285 Noise in Physical Systems and 1/f Fluctuations (St. Louis, MO 1993)	93-72575	1-56396-270-5
No. 286 Ordering Disorder: Prospect and Retrospect in Condensed Matter Physics: Proceedings of the Indo-U.S. Workshop (Hyderabad, India 1993)	93-072549	1-56396-255-1
No. 287 Production and Neutralization of Negative Ions and Beams: Sixth International Symposium (Upton, NY 1992)	93-72821	1-56396-103-2
No. 288 Laser Ablation: Mechanismas and Applications-II: Second International Conference (Knoxville, TN 1993)	93-73040	1-56396-226-8
No. 289 Radio Frequency Power in Plasmas: Tenth Topical Conference (Boston, MA 1993)	93-72964	1-56396-264-0
No. 290 Laser Spectroscopy: XIth International Conference (Hot Springs, VA 1993)	93-73050	1-56396-262-4
No. 291 Prairie View Summer Science Academy (Prairie View, TX 1992)	93-73081	1-56396-133-4
No. 292 Stability of Particle Motion in Storage Rings (Upton, NY 1992)	93-73534	1-56396-225-X
No. 293 Polarized Ion Sources and Polarized Gas Targets (Madison, WI 1993)	93-74102	1-56396-220-9
No. 294 High-Energy Solar Phenomena: A New Era of Spacecraft Measurements (Waterville Valley, NH 1993)	93-74147	1-56396-291-8
No. 295 The Physics of Electronic and Atomic Collisions: XVIII International Conference (Aarhus, Denmark, 1993)	93-74103	1-56396-290-X
No. 296 The Chaos Paradigm: Developments an Applications in Engineering and Science (Mystic, CT 1993)	93-74146	1-56396-254-3
No. 297 Computational Accelerator Physics (Los Alamos, NM 1993)	93-74205	1-56396-222-5

Title	L.C. Number	ISBN
No. 298 Ultrafast Reaction Dynamics and Solvent Effects (Royaumont, France 1993)	93-074354	1-56396-280-2
No. 299 Dense Z-Pinches: Third International Conference (London, 1993)	93-074569	1-56396-297-7
No. 300 Discovery of Weak Neutral Currents: The Weak Interaction Before and After (Santa Monica, CA 1993)	94-70515	1-56396-306-X
No. 301 Eleventh Symposium Space Nuclear Power and Propulsion (3 Vols.) (Albuquerque, NM 1994)	92-75162	1-56396-305-1 (set) 156396-301-9 (pbk. set)
No. 302 Lepton and Photon Interactions/ XVI International Symposium (Ithaca, NY 1993)	94-70079	1-56396-106-7
No. 303 Slow Positron Beam Techniques for Solids and Surfaces Fifth International Workshop (Jackson Hole, WY 1992)	94-71036	1-56396-267-5
No. 304 The Second Compton Symposium (College Park, MD 1993)	94-70742	1-56396-261-6
No. 305 Stress-Induced Phenomena in Metallization Second International Workshop (Austin, TX 1993)	94-70650	1-56396-251-9
No. 306 12th NREL Photovoltaic Program Review (Denver, CO 1993)	94-70748	1-56396-315-9
No. 307 Gamma-Ray Bursts Second Workshop (Huntsville, AL 1993)	94-71317	1-56396-336-1
No. 308 The Evolution of X-Ray Binaries (College Park, MD 1993)	94-76853	1-56396-329-9
No. 309 High-Pressure Science and Technology—1993 (Colorado Springs, CO 1993)	93-72821	1-56396-219-5 (set)
No. 310 Analysis of Interplanetary Dust (Houston, TX 1993)	94-71292	1-56396-341-8
No. 311 Physics of High Energy Particles in Toroidal Systems (Irvine, CA 1993)	94-72098	1-56396-364-7
No. 312 Molecules and Grains in Space (Mont Sainte-Odile, France 1993)	94-72615	1-56396-355-8
No. 313 The Soft X-Ray Cosmos ROSAT Science Symposium (College Park, MD 1993)	94-72499	1-56396-327-2

Title	L.C. Number	ISBN
No. 314 Advances in Plasma Physics Thomas H. Stix Symposium (Princeton, NJ 1992)	94-72721	1-56396-372-8
No. 315 Orbit Correction and Analysis in Circular Accelerators (Upton, NY 1993)	94-72257	1-56396-373-6
No. 316 Thirteenth International Conference on Thermoelectrics (Kansas City, Missouri 1994)	95-75634	1-56396-444-9
No. 317 Fifth Mexican School of Particles and Fields (Guanajuato, Mexico 1992)	94-72720	1-56396-378-7
No. 318 Laser Interaction and Related Plasma Phenomena 11th International Workshop (Monterey, CA 1993)	94-78097	1-56396-324-8
No. 319 Beam Instrumentation Workshop (Santa Fe, NM 1993)	94-78279	1-56396-389-2
No. 320 Basic Space Science (Lagos, Nigeria 1993)	94-79350	1-56396-328-0
No. 321 The First NREL Conference on Thermophotovoltaic Generation of Electricity (Copper Mountain, CO 1994)	94-72792	1-56396-353-1
No. 322 Atomic Processes in Plasmas Ninth APS Topical Conference (San Antonio, TX)	94-72923	1-56396-411-2
No. 323 Atomic Physics 14 Fourteenth International Conference on Atomic Physics (Boulder, CO 1994)	94-73219	1-56396-348-5
No. 324 Twelfth Symposium on Space Nuclear Power and Propulsion (Albuquerque, NM 1995)	94-73603	1-56396-427-9
No. 325 Conference on NASA Centers for Commercial Development of Space (Albuquerque, NM 1995)	94-73604	1-56396-431-7
No. 326 Accelerator Physics at the Superconducting Super Collider (Dallas, TX 1992-1993)	94-73609	1-56396-354-X
No. 327 Nuclei in the Cosmos III Third International Symposium on Nuclear Astrophysics (Assergi, Italy 1994)	95-75492	1-56396-436-8
No. 328 Spectral Line Shapes, Volume 8 12th ICSLS (Toronto, Canada 1994)	94-74309	1-56396-326-4
No. 329 Resonance Ionization Spectroscopy 1994 Seventh International Symposium (Bernkastel-Kues, Germany 1994)	95-75077	1-56396-437-6

Title	L.C. Number	ISBN
No. 330 E.C.C.C. 1 Computational Chemistry F.E.C.S. Conference (Nancy, France 1994)	95-75843	1-56396-457-0
No. 331 Non-Neutral Plasma Physics II (Berkeley, CA 1994)	95-79630	1-56396-441-4
No. 332 X-Ray Lasers 1994 Fourth International Colloquium (Williamsburg, VA 1994)	95-76067	1-56396-375-2
No. 333 Beam Instrumentation Workshop (Vancouver, B. C., Canada 1994)	95-79635	1-56396-352-3
No. 334 Few-Body Problems in Physics (Williamsburg, VA 1994)	95-76481	1-56396-325-6
No. 335 Advanced Accelerator Concepts (Fontana, WI 1994)	95-78225	1-56396-476-7 (set) 1-56396-474-0 (Book) 1-56396-475-9 (CD-Rom)
No. 336 Dark Matter (College Park, MD 1994)	95-76538	1-56396-438-4
No. 337 Pulsed RF Sources for Linear Colliders (Montauk, NY 1994)	95-76814	1-56396-408-2
No. 338 Intersections Between Particle and Nuclear Physics 5th Conference (St. Petersburg, FL 1994)	95-77076	1-56396-335-3
No. 339 Polarization Phenomena in Nuclear Physics Eighth International Symposium (Bloomington, IN 1994)	95-77216	1-56396-482-1
No. 340 Strangeness in Hadronic Matter (Tucson, AZ 1995)	95-77477	1-56396-489-9
No. 341 Volatiles in the Earth and Solar System (Pasadena, CA 1994)	95-77911	1-56396-409-0
No. 342 CAM -94 Physics Meeting (Cacun, Mexico 1994)	95-77851	1-56396-491-0
No. 343 High Energy Spin Physics Eleventh International Symposium (Bloomington, IN 1994)	95-78431	1-56396-374-4
No. 344 Nonlinear Dynamics in Particle Accelerators: Theory and Experiments (Arcidosso, Italy 1994)	95-78135	1-56396-446-5
No. 345 International Conference on Plasma Physics ICPP 1994 (Foz do Iguaçu, Brazil 1994)	95-78438	1-56396-496-1

Title	L.C. Number	ISBN
No. 346 International Conference on Accelerator-Driven Transmutation Technologies and Applications (Las Vegas, NV 1994)	95-78691	1-56396-505-4
No. 347 Atomic Collisions: A Symposium in Honor of Christopher Bottcher (1945-1993) (Oak Ridge, TN 1994)	95-78689	1-56396-322-1
No. 348 Unveiling the Cosmic Infrared Background (College Park, MD, 1995)	95-83477	1-56396-508-9
No. 349 Workshop on the Tau/Charm Factory (Argonne, IL, 1995)	95-81467	1-56396-523-2
No. 350 International Symposium on Vector Boson Self-Interactions (Los Angeles, CA 1995)	95-79865	1-56396-520-8
No. 351 The Physics of Beams Andrew Sessler Symposium (Los Angeles, CA 1993)	95-80479	1-56396-376-0
No. 352 Physics Potential and Development of $\mu^+ \mu^-$ Colliders: Second Workshop (Sausalito, CA 1994)	95-81413	1-56396-506-2
No. 353 13th NREL Photovoltaic Program Review (Lakewood, CO 1995)	95-80662	1-56396-510-0
No. 354 Organic Coatings (Paris, France, 1995)	96-83019	1-56396-535-6
No. 355 Eleventh Topical Conference on Radio Frequency Power in Plasmas (Palm Springs, CA 1995)	95-80867	1-56396-536-4
No. 356 The Future of Accelerator Physics (Austin, TX 1994)	96-83292	1-56396-541-0
No. 357 10th Topical Workshop on Proton-Antiproton Collider Physics (Batavia, IL 1995)	95-83078	1-56396-543-7
No. 358 The Second NREL Conference on Thermophotovoltaic Generation of Electricity	95-83335	1-56396-509-7
No. 359 Workshops and Particles and Fields and Phenomenology of Fundamental Interactions (Puebla, Mexico 1995)	96-85996	1-56396-548-8
No. 360 The Physics of Electronic and Atomic Collisions XIX International Conference (Whistler, Canada, 1995)	95-83671	1-56396-440-6
No. 361 Space Technology and Applications International Forum (Albuquerque, NM 1996)	95-83440	1-56396-568-2
No. 362 Two-Center Effects in Ion-Atom Collisions (Lincoln, NE 1994)	96-83379	1-56396-342-6

	Title	L.C. Number	ISBN
No. 363	Phenomena in Ionized Gases XXII ICPIG (Hoboken, NJ, 1995)	96-83294	1-56396-550-X
No. 364	Fast Elementary Processes in Chemical and Biological Systems (Villeneuve d'Ascq, France, 1995)	96-83624	1-56396-564-X
No. 365	Latin-American School of Physics XXX ELAF Group Theory and Its Applications (México City, México, 1995)	96-83489	1-56396-567-4
No. 366	High Velocity Neutron Stars and Gamma-Ray Bursts (La Jolla, CA 1995)	96-84067	1-56396-593-3
No. 367	Micro Bunches Workshop (Upton, NY, 1995)	96-83482	1-56396-555-0
No. 368	Acoustic Particle Velocity Sensors: Design, Performance and Applications (Mystic, CT, 1995)	96-83548	1-56396-549-6
No. 369	Laser Interaction and Related Plasma Phenomena (Osaka, Japan 1995)	96-85009	1-56396-445-7
No. 370	Shock Compression of Condensed Matter-1995 (Seattle, WA 1995)	96-84595	1-56396-566-6
No. 371	Sixth Quantum 1/f Noise and Other Low Frequency Fluctuations in Electronic Devices Symposium (St. Louis, MO, 1994)	96-84200	1-56396-410-4
No. 372	Beam Dynamics and Technology Issues for + - Colliders 9th Advanced ICFA Beam Dynamics Workshop (Montauk, NY, 1995)	96-84189	1-56396-554-2
No. 373	Stress-Induced Phenomena in Metallization (Palo Alto, CA 1995)	96-84949	1-56396-439-2
No. 374	High Energy Solar Physics (Greenbelt, MD 1995)	96-84513	1-56396-542-9
No. 375	Chaotic, Fractal, and Nonlinear Signal Processing (Mystic, CT 1995)	96-85356	1-56396-443-0
No. 376	Chaos and the Changing Nature of Science and Medicine: An Introduction (Mobile, AL 1995)	96-85220	1-56396-442-2
No. 377	Space Charge Dominated Beams and Applications of High Brightness Beams (Bloomington, IN 1995)	96-85165	1-56396-625-7
No. 378	Surfaces, Vacuum, and Their Applications (Cancun, Mexico 1994)	96-85594	1-56396-418-X
No. 379	Physical Origin of Homochirality in Life (Santa Monica, CA 1995)	96-86631	1-56396-507-0

	Title	L.C. Number	ISBN
No. 380	Production and Neutralization of Negative Ions and Beams / Production and Application of Light Negative Ions (Upton, NY 1995)	96-86435	1-56396-565-8
No. 381	Atomic Processes in Plasmas (San Francisco, CA 1996)	96-86304	1-56396-552-6
No. 382	Solar Wind Eight (Dana Point, CA 1995)	96-86447	1-56396-551-8
No. 383	Workshop on the Earth's Trapped Particle Environment (Taos, NM 1994)	96-86619	1-56396-540-2
No. 384	Gamma-Ray Bursts (Huntsville, AL 1995)	96-79458	1-56396-685-9
No. 385	Robotic Exploration Close to the Sun: Scientific Basis (Marlboro, MA 1996)	96-79560	1-56396-618-2
No. 386	Spectral Line Shapes, Volume 9 13th ICSLS (Firenze, Italy 1996)		1-56396-656-5
No. 387	Space Technology and Applications International Forum (Albuquerque, NM 1997)	96-80254	1-56396-679-4 (Case set) 1-56396-691-3 (Paper set)
No. 388	Resonance Ionization Spectroscopy 1996 Eighth International Symposium (State College, PA 1996)	96-80324	1-56396-611-5
No. 389	X-Ray and Inner-Shell Processes 17th International Conference (Hamburg, Germany 1996)	96-80388	1-56396-563-1
No. 390	Beam Instrumentation Proceedings of the Seventh Workshop (Argonne, IL 1996)	97-70568	1-56396-612-3
No. 391	Computational Accelerator Physics (Williamsburg, VA 1996)	97-70181	1-56396-671-9
No. 392	Applications of Accelerators in Research and Industry: Proceedings of the Fourteenth International Conference (Denton, TX 1996)	97-71846	1-56396-652-2
No. 393	Star Formation Near and Far Seventh Astrophysics Conference (College Park, MD 1996)	97-71978	1-56396-678-6
No. 394	NREL/SNL Photovoltaics Program Review Proceedings of the 14th Conference— A Joint Meeting (Lakewood, CO 1996)	97-72645	1-56396-687-5

	Title	L.C. Number	ISBN
No. 395	Nonlinear and Collective Phenomena in Beam Physics (Arcidosso, Italy 1996)	97-72970	1-56396-668-9
No. 396	New Modes of Particle Acceleration— Techniques and Sources (Santa Barbara, CA 1996)	97-72977	1-56396-728-6
No. 397	Future High Energy Colliders (Santa Barbara, CA 1997)	97-73333	1-56396-729-4
No. 398	Advanced Accelerator Colliders Seventh Workshop (Lake Tahoe, CA 1996)	97-72788	1-56396-697-2 (set) 1-56396-727-8 (cloth) 1-56396-726-X (CD-Rom)
No. 399	The Changing Role of Physics Departments (College Park, MD 1996)		1-56396-698-0
No. 400	High Energy Physics First Latin Symposium (Yucatan, México 1996)	97-73971	1-56396-686-7
No. 401	Thermophotovoltaic Generation of Electricity Third NREL Conference (Colorado Springs, CO 1997)	97-74374	1-56396-734-0
No. 403	Radio Frequency Power in Plasmas 12th Topical Conference (Savannah, GA 1997)	97-74472	1-56396-709-X
No. 404	Future Generations Photovoltaic Technologies First NREL Conference (Denver, CO 1997)	97-74386	1-56396-704-9
No. 405	Beam Stability and Nonlinear Dynamics (Santa Barbara, CA 1996)	97-74676	1-56396-731-6